First Science Investigations

Norman G. Furnell

Blackwell Education

Illustrated by Lorraine White

Typeset in Rockwell light and bold
by VAP Publishing Services, Kidlington, Oxford

Paper is 90 gsm Ashtead Opaque

Printed and bound in Great Britain
by Dotesios (Printers) Ltd, Bradford on Avon

Contents

I hear – I forget; I see – I remember; I do – I understand.
Ancient Chinese proverb

Preface

The investigations in the course introduce children to science as a way of learning through experiment. The work is not based on science as a body of factual knowledge and neither the pupils nor the teacher need to have any previous scientific or technical skills. Children working in small groups can carry out the activities by following simply written and clearly illustrated instructions provided as photocopiable master sheets. Detailed teacher's notes accompany each activity.

Acknowledgements

My special thanks to the children and staff who participated in the trials at the following schools: Paddock Wood County Primary School, Paddock Wood, Kent; Little Houghton Primary School, Little Houghton, Northants.; and Longmead Junior School, Tonbridge, Kent; and for contributions from Cogenhoe Primary and Earls Barton Junior Schools, both in Northants.

Geoff Welford kindly reviewed the text and I am most grateful for his thoroughness and the many valued comments and suggestions he has made.

Introduction

How to use the book

How will science fit into my work?
See pages 2 and 3

How do I start off?
See pages 3, 4 and 14

How can the course be used?
See pages 3–4

What will be achieved?
See pages 14–16

What age range is it for?
See pages 1–2

First science investigations

What investigations are there?
See page iii

How is the work organised?
See pages 3–4

What about written work?
See pages 5–7

What about evaluation?
See pages 9–16

An outline of how the investigations may be used, together with general information, is given on pp. 3–4. A description of the nature of each investigation can be found in the list of investigations and more detailed information is given in the teacher's sheets and children's instruction sheets.

The word *science* has several definitions. It can refer to a body of knowledge such as chemistry or physics or it can mean a system of study involving experimentation requiring the process skills now being emphasised in GCSE science courses.

First Science Investigations is based on science as a system of study and is not reliant on a body of knowledge for success.

The 20 investigations in the course are designed as classroom-based activities which can be used to help children learn and develop the skills and abilities used in the process of science. By carrying out such investigations children learn to handle experimental variables with confidence and also acquire a deeper understanding of the phenomena under investigation.

Age-range of target pupils

The age-range for which this work is suitable is less important than the level of conceptual development of children (see *National Curriculum Science Working Group (Interim Report)*, D.E.S., 1987 or *Science 5–13, With Objectives in Mind*, Macdonald Educational, 1974).

PRIMARY CHILDREN

Trials have shown the material to be suitable for third- and fourth-year juniors. Restricting the work to top juniors initially will simplify starting off for the first time. Children able to carry out division using a calculator will have an advantage.

SECONDARY CHILDREN

All of the investigations are suitable for first- and second-year pupils. For less able third-year pupils investigations 10–20 offer a wide variety of stimulating activities. Investigations 15–20 are suitable for groups who are practised in variable-based investigations similar to those included earlier (see 'Choosing investigations' below).

Integration of the investigations

Investigations are based on science as a systematic study with broad overlaps into art, craft, mathematics and music. Many of the activities integrate well into thematic work such as Discovery, Myself, Shape and Pattern. The variety of tasks and inclusion of extension work enables them to be used with a wide age and ability range.

Choosing investigations to use

Before choosing an investigation, read the teacher's notes to check its suitability and the equipment and materials required. Copies should be made of the pupils' sheets and the table sheets provided.

The order of the investigations in the text is an indication of their degree of difficulty.

1–7
These activities require no special scientific apparatus or skills. The extension work included gives extra scope to some of the activities.

8–14
This section introduces work with animals, chemicals (in 'Crystal garden'), and some specialist apparatus such as forcemeters. Several of the activities require a basic manual dexterity and are suitable for developing manipulative skills. Calculations include finding averages and division.

15–20

These activities involve the use of hot water and tools such as a craft knife, and require reasonably developed manipulative skills. They are likely to take several hours to complete and could form the basis of quite lengthy investigations. Calculations include finding averages, division, and multiplication with numbers to three decimal places.

This is a general guide only; some children who are competent at arithmetic, for example, may find some later activities more to their liking than earlier ones, and so on.

Using the investigations

There are two ways that the material can be used.

1 As part of general class activities in the primary classroom

The work has been successfully used in trials with children working in pairs on different investigations. The most difficult part is starting off for the first time!

From the point of organisation it is sensible to have groups doing a variety of other tasks as well as science. If activities such as painting, reading, mathematical games and language work are included it means that the teacher can limit the numbers starting the science work.

Investigations 1–7 are the most suitable for groups to start with, as they require no special skills or scientific apparatus. Once teachers are familiar with the material, they can try the investigations with younger children who may need help at first with filling in tables, finding averages and writing about their work.

Having been given the equipment and instructions, a small group can get on and do the work, each child producing their own report. Examples of bar charts and other children's work can help here but some children have a tendency to copy out work on display!

Once one group of children gains confidence in carrying out the investigations, it has a positive effect on the rest of the class.

2 As a class science lesson

The investigations can be carried out as a circus of activities. Equipment permitting, a class conducting a maximum of four or five different activities is easier to manage. It is impractical to have large numbers of children carrying out survey work at the same time.

Teacher's sheets

Each investigation has accompanying guidelines for the teacher. The purpose of the work is explained and background information from the trials is included. Suitable methods are suggested for activities where difficulties might arise. There is a list of the apparatus required as well as recommended brands and suppliers of specific equipment.

Children's instruction sheets

The aim has been to make the sheets as helpful as possible without over-direction, so that children working in small groups can incorporate their own ideas and carry out the investigations with minimal assistance. The illustrations serve as 'visual equipment lists' and often indicate a possible procedure so that even non-readers can get some idea of the tasks involved. The treatment of results, and method for writing up, are outlined.

The instructions for carrying out investigations are deliberately limited to avoid prescription. It is hoped that once children have completed a few investigations and become more confident they will want to experiment and introduce their own ideas. Even though alternative methods are not always perfect, if they are safe they should be encouraged. Directions need only be given when groups cannot proceed at all – a bit of struggling on their part may lead to an 'I do, I understand' outcome!

Terms used in investigations

It will be helpful for children if they are familiar with the terms listed below before starting their investigations.

1–7
- bar chart
- conclusion
- experiment

- results
- survey (asking people their opinion)
- table (for results)

8–14
- cubic centimetres (cm^3)
- forcemeter
- grams (g)
- kilograms (kg)
- measuring cylinder
- newtons (N)
- percentages (%)
- scoring (as awarding points)
- thermometer (to measure degrees Celsius)

15–20
- additives
- antiseptic
- Celsius (degrees)
- cooling curve (see 'Cooling down – covering up')
- insulator

Examples of work

The children's work shown here has been taken from our trials of the investigations. The examples illustrate two of the ways that investigations can be written up and also provide a good example of what is meant by fair testing. The teacher's notes to some of the investigations include further examples of children's work.

WRITING UP

This may be an unstructured report where the conclusion is simply a comment about the result (see the figure below, from 'Righting wrongs').

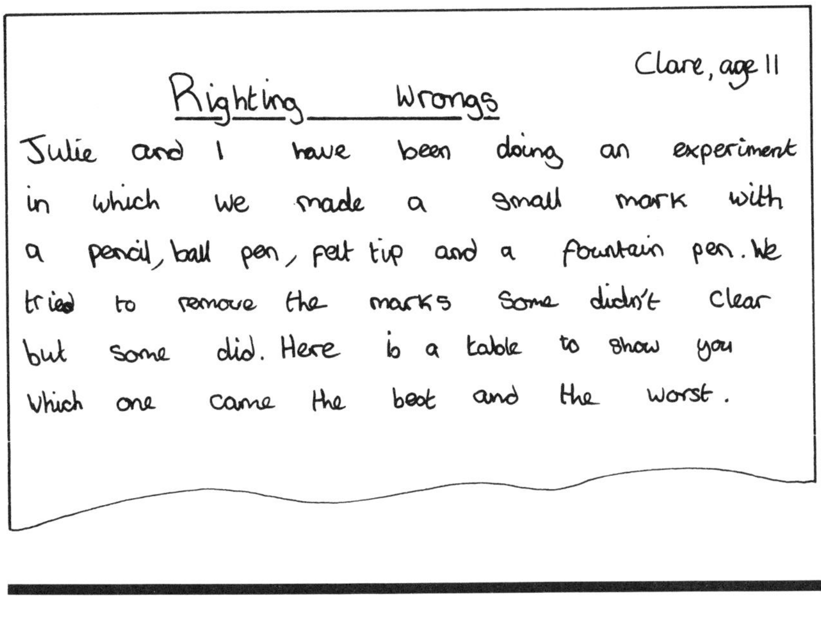

Clare, age 11

Righting Wrongs

Julie and I have been doing an experiment in which we made a small mark with a pencil, ball pen, felt tip and a fountain pen. We tried to remove the marks some didn't clear but some did. Here is a table to show you which one came the best and the worst.

Method	Score out of 10				Average score
	Pencil	Ball pen	Felt tip	Ink	
Ink rubber	7	3	2	6	4·5
Typists rubber	8	2	9	4	5·75
Natural rubber	9	1	1	3	3·5
Tipp-Ex paper	9	9	9	9	9
Tipp-Ex pen	8	9	8	10	8·75
Eraser pen	1	4	2	0	1·75

The tipp-ex paper was the best from the average score.
The eraser pen was the worst it made a dirty mark.

Alternatively it may be a structured report presented under the following headings: Aim, Apparatus, Method, Results and Conclusion (see the figure below, from 'Orange squash').

Gina, age 10.

Orange squash

Aim
    To find the best Orange Squash.

Apparatus
    Orange Squashes
    measuring cylinders
    plastic cups
    marker pens.

method
    we made up some Orange drinks and measured
    how much Orange Squash we added.
    We tested the drinks to get them the right
    strength and then we did a taste test.

Result

Name of squash	Squash costs				
	Price (p)	Volume (cm³)	Cost per cm³ (p/cm³)	Squash added (cm³)	Cost per drink (p)
Star Squash	37p	1000CC	0.037	10CC	0.37
all Juice	57p	1000CC	0.057	10CC	0.57
Costa Squash	52p	1000CC	0.052	20CC	1.04
Crusha	65	700CC	0.089	30CC	2.67
Appeel	45p	1000CC	0.045	40CC	0.45
Squasha	69p	900CC	0.077	10CC	0.77

Squash	Taste /6	Cost per drink /6	Total score
Star Squash	4	6	10
All Juice	6	4	10
Costa Squash	3	2	5
Crusha	2	1	3
Appeel	5	5	10
Squasha	1	3	4

Conclusion.

the best ones were Star Squash all Juice and Appeel. Costa Sqash was fourth Squasha. was fifth and Crusha was Last. the drinks must be the right strength to taste right

FAIR TESTING

It is important that testing be carried out correctly. Two examples of work from trials are shown in figures from 'Making-up faces'. A fair test is shown here. The make-up is being investigated and this is different for each face.

By contrast, an unfair test is shown which includes different iris colours, hair, a hat and jewellery. This means things are being changed other than the make-up – that is, different variables have been introduced which may affect the result.

Work evaluation

AN ASSESSMENT SCHEME

As work progresses children's achievement can be
assessed using a scheme like the one below which is
based on the criteria listed in the Objectives (page 15).
The six sections included cover the different scientific
skills and abilities being evaluated:

Planning

		Marks
a	Attempt made to produce a method	1
b	Bases plan on a fair test	2
c	Produces a working method	3
d	Selects suitable apparatus to use	4
e	Modifies method in light of any difficulties encountered	5

Practical work

One mark for each of the following abilities:
- Can work to instructions provided
- Uses apparatus safely
- Makes careful measurements/observations
- Can solve minor difficulties without help
- Is able to use own initiative

Report evaluation
Description of investigation

		Marks
a	Attempt made to explain the method	1
b	Generally correct account of method	2
c	Reference made to apparatus used	3
d	Purpose of investigation indicated	4
e	Purpose of investigation correctly stated, method clearly expressed	5

Results

		Marks
a	Few results obtained	1
b	Results more or less complete	2
c	Totals found/averages taken	3
d	Results presented graphically	4
e	Good graphical presentation (e.g. suitable scales, correct labels)	5

Conclusion

		Marks
a	Attempt made to say what the results show	1
b	Generally correct conclusion made	2
c	What the results show clearly identified	3
d	An interpretation based on the results obtained	4
e	Correct conclusion and interpretation, improvements to investigation suggested	5

Refinements

One mark for each of the following actions:
- Double-checks any unusual readings
- Shows understanding of the limitations of results
- Takes precautions to improve safety or accuracy
- Explains how minor difficulties were overcome
- Having completed an investigation, suggests further actions

For the 'Practical work' and 'Refinements' sections, 1 mark is suggested for each action or ability included. For the other sections, marks are awarded on a 1–5 scale. In these sections each step includes a requirement in addition to the previous one. For example, under 'Results', 'Totals found/Averages taken' assumes that the previous step, 'Results more or less complete', has been achieved, and so on.

The 'Planning' section applies to investigations the children design for themselves which will arise with extension work or their own follow-up work.

The 'Practical work' section can be used by observing the children as they carry out an investigation.

Often in children's work some points are missed and other points are out of the order given in the scheme. This means that a degree of flexibility is needed in its application and it should not be regarded as being unadaptable.

Not all of the elements of the scheme are applicable to some of the investigations; for example, graphical presentation is not always appropriate. In such cases the marks awarded should be adjusted accordingly.

Help

Some children are able to carry out their own ideas with little or no help while others need constant guidance. The teacher may want to take account of this for assessment purposes; the example below is given assuming marks out of 5:

	Marks
Occasional help (per section)	−1
Frequent help	−2
Considerable help	−3

USING THE ASSESSMENT SCHEME

This work from the investigation 'Which soap?' (see page 85) has been evaluated using the sections for 'Report evaluation' and 'Planning':

Which Soap!

We were asked to test different soaps out. The soaps that we tested were Lux, Household soap, Shield, Lemon Fresh and Palmolive. We had to test all the soaps out on their Cleaning Power, Appeal, and their costs. We were looking for the best value soap.

First of all we tested the soaps cleaning power.

Samantha, Anna, Judith and I had to make our fingers messy and then clean them off. Kurt was timekeeper. One person was grime, one was shoe polish, one was ink, and one was a permanent marker. I was a permanent marker. We each had 1 minute to get the mark off one finger with each soap. We used one soap for each finger.

These are our results:

Brand of soap	Scores out of five for cleaning					
	Felt Pen	Ink	Grime	Polish	Average	/5
Lux	1/5	1/5	5/5	5/5	3	5/5
Household	0/5	2/5	5/5	5/5	3	5/5
Shield	0/5	2/5	5/5	5/5	3	5/5
Lemon Fresh	0/5	2/5	5/5	5/5	3	5/5
Palmolive	0/5	0/5	5/5	5/5	2.5	1/5

We found that Lux, Household soap, Shield and Lemon Fresh all had an overall score of 12 and an average score of 3.

Palmolive had an overall score of 10 and an average score of 2.5.

Lux, Household soap, Shield and Lemon Fresh came first, and Palmolive came last.

We then tested the soaps appeal.

These are some of the things that we thought made soaps nice to use,

The scent: Some soaps do not smell very nice when they have been used.

The texture: Some soaps go rough after a while.

The shape: Some soaps you cannot fit your hands round them, which makes them hard to use.

The colour: Some soaps colours can put people off buying them.

The wrapper: Some people won't buy soaps with boring wrappers

The Name: Something like household soap will not attract many people.

| Brand of soap | Scores out of five. | | | | Average | /5 |
	Shape	Smell	Wrapper	Colour.		
Household.	1/5	2/5	0/5	0/5	0.75	1/5
Lux.	4/5	5/5	4/5	5/5	4.5	5/5
Shield	3/5	2/5	4/5	2/5	2.75	3/5
Lemon Fresh	3/5	4/5	3/5	4/5	3.5	4/5
Palmolive.	3/5	2/5	1/5	2/5	2.0	2/5

Then we looked at the soaps cost.

Household soap was the best value, then Shield, then came Palmolive and Lux, then last came Lemon Fresh.

Brand of soap.	Price (p)	mass of soap (g)	Cost per gram (p.g)	/5
Lemon Fresh.	22p	125g	0.176	1/5
Lux.	19p	125g	0.152	3/5
Shield.	18p	125g	0.144	4/5
Palmolive.	19p	125g	0.152	3/5
Household soap.	23p	163g	0.1411042	5/5

Final Score.

These are the final results:

| Brand of Soap | Scores out of five. | | | Aggregate Score | Average Score. |
	Price.	Cleaning Power.	Appeal.		
Lux.	3/5	5/5	5/5	13	4.3
Household	5/5	5/5	1/5	11	3.6
Palmolive	3/5	1/5	2/5	6	2
Shield	4/5	5/5	3/5	12	4
Lemon Fresh.	1/5	5/5	4/5	10	3.3

We discovered that Lux was the best soap, Shield came next, then Household soap, then Lemon Fresh and last came Palmolive.

Lux scored five out of five on appeal and cleaning power and scored 3 on price. I agree with Lux being the best soap.

Extension Work

Get a young child a child of about 12 and an adult. Weigh a certain brand of soap. Then get the people to wash their hands with a bar of soap each, for about 3 minutes see how much of the soap has been used by weighing it again If the 1st time the bar weighed 125g and the second time it weighed 120g Then the soap could be used for 25 washes. This test is fair because the soap is being tested on Children and adults

Evaluation

Description of investigation: Scores 5 – the purpose of the investigation is correctly stated, the apparatus used is described and the method is clearly expressed.

Results: Scores 3 – tables showing averages. Note the last entry under 'Cost per gram' in the third table of results. 0.141 would be more appropriate than the figure quoted to seven decimal places!

Conclusion: Scores 3 – what the results show has been identified.

Refinements: Scores 0 – any actions such as taking precautions to improve accuracy or suggesting further tests are not reported.

Planning: This section applies to the extension work in the report. Scores 3 – a working method has been described.

Total score = 14 marks

Further evaluations using the scheme are included in the teacher's sheets for investigations 1–7.

SELF EVALUATION

Children can be given a question sheet to answer after completing an investigation. Questions might include:

a Did you enjoy the work?   _____
b Have you understood the work?   _____
c Have you checked for spelling mistakes in
   your report?   _____
d Have you included a diagram in your
   method?   _____
e Does your conclusion say what you found
   out from the experiment?   _____

In checking their work children can become much more critical of its content and presentation, and many make a real effort to improve their standards.

CORRECT ANSWERS

Provided that experimental results are obtained from accurate observations made by the experimenter, they are correct answers. Even if the observed results are not what was expected it does not mean that the work was not carried out properly. For example, in the 'Orange squash' investigation, two groups could select different brands as the best buy due to differences in personal choice. In both cases fair tests could have been carried out in a sound manner.

Practical considerations

For long-term use, children's instruction sheets can be stuck on hardboard cut to A4 size and covered in self-adhesive covering film.

The display graphs in the final section (pages 120–124) give suitable scales and axes. They show the form that the results might take but should not be taken as correct answers. It may help children if they can refer to these during the activity.

Result tables are illustrated on the children's instruction sheets. Master sheets of the tables are included in the final section (pages 111–119).

For the investigations 'Making-up faces', 'Orange squash' and 'Penny whistle', children should carry out the activities individually, even when working in groups.

It is advisable to keep the equipment for an investigation, particularly tools and fragile apparatus such as thermometers, in a tray.

Use of calculators

About half of the activities involve finding averages, multiplication and division. These tasks will be greatly simplified if a calculator can be used with confidence.

Answers should not be given to more than two decimal places (with the exception of values for cost per cm^3 and cost per gram in 'Which soap?' and 'Orange squash').

Aims and objectives

The intention has been to provide activities for younger children which involve a range of skills based on the aims and assessment objectives in the GCSE National Criteria for science subjects.

AIMS

To maintain and encourage the natural curiosity of young children for the world around them.

To help develop in children
• the idea of science as a system of study,
• an understanding of the nature of the science process,

- the basic manipulative and communicative skills needed to carry out science investigations,
- skills of observation, and
- an awareness of the importance of accurate experimental work.

To help children develop the abilities to
- record and tabulate results,
- present data graphically and in tabular form,
- interpret and evaluate their recorded data, and
- carry out experiments in a safe manner.

To provide a foundation of skills and processes for future science work.

OBJECTIVES

Experimental work
Children should be starting to:
- plan and carry out simple experimental investigations to test out ideas,
- follow instructions to carry out practical investigations,
- choose suitable apparatus for an activity from a given selection,
- handle and use basic laboratory apparatus in a safe manner,
- make and record observations carefully,
- record the results of experiments accurately and clearly,
- present results in the form of bar charts and line graphs,
- analyse and interpret the results of their experiments,
- draw conclusions from their experiments,
- criticise the design of experiments they have tried, and
- produce reports of work carried out.

Recall/knowledge
Children should be able to recall
- the steps in the science process and
- simple fundamental and derived units they have used.

Understanding
Children should be starting to
- present, use and interpret the results of investigations in numerical, written or graphical form and to change them from one form to another, and
- carry out simple arithmetic calculations according to given instructions.

Application analysis and evaluation

Children should be starting to

- draw conclusions and make generalisations and predictions from observations and experimental results, and
- use their curiosity about their environment in a more effective manner.

Summary

Cards of different fluorescent colours are placed outside. The attractiveness of the colours to insects is determined.

Equipment

- A4 size fluorescent coloured cards (red, orange, yellow, green, mauve)
- Clock/watch
- Flat area outside (e.g. a grass verge away from traffic)

For extension work
- Normal (non-fluorescent) coloured cards
- Insect identification book

Notes

On a warm sunny day the air can be full of flying insects like aphids and hover flies. This investigation sets out to see if they are attracted to some colours more than others. If the coloured cards are placed in an open area such as a grassy verge, insects will land in surprisingly large numbers in less than a minute. In practice the orange and yellow attract more insects than other colours do.

During trials some children wanted to put the cards near flowers and even on bushes (where they thought the insects would be!). Strictly, this is introducing bias. In what are called *sampling techniques*, sites should be picked at random and not specially selected. (See, for example, *Statistics for Biologists*, R.C. Campbell, C.U.P., 1967.)

Groups may want to try out various locations. Ensure that they work safely – away from traffic and where they can be seen and supervised.

The season and weather conditions are important. Ideal conditions will be on a day when butterflies are abundant. Cold days, especially at times of frost, are unlikely to be successful. If the weather is not ideal and few flies are in the air, it may help to disturb insects by walking through nearby grassy areas.

Fluorescent coloured card is obtainable from stationers either in assorted packs or as A3 sheets. Unless treated carefully the cards soon become damp, marked and creased. In windy weather the cards blow away!

Children's work

<u>Insects landing</u>
We got some coloured cards and we went outside. We put the coloured paper on the ground and we away and then we came back. When we saw some Insects we put a tick in a chart.
We put the cards in three different places and put the results on a chart

COLOUR OF CARD	NUMBER OF INSECTS			
	1	2	3	AVERAGE
RED	2	0	h	2
ORANGE	1	3	5	3
YELLOW	3	4	10	5.7
GREEN	11	7	11	9.7
MAUVE	2	0	5	2.3

<u>conclusion</u>
The most popular colour was green because they think its a leaves. The colour red was the least popular.

Report evaluation

Description of investigation: Scores 3 – generally correct account of method and reference made to apparatus used.

Results: Scores 3 – table showing averages.

Conclusion: Scores 3 – what the results show has been identified but the idea about the leaves is what the investigators think the insects think! This is not based on the results obtained, it is merely an assumption.

Total score = 9 marks

Insects landing

The air outside is full of flying insects which
land on flowers, leaves or animals. What
makes an insect decide to land?

Your job is to see if the colour of an object
makes any difference to the insects that land
on it.

You need
- Coloured cards
- Clock or watch

Do this test on a nice day when insects are
flying around. Ask permission before you go
outside.

Shuffle the cards, then put them down on a flat
grassy area. See which colour the insects land
on most.

If you do not see insects on the cards quite
quickly, try walking through the grass around
them.

It is up to you how long you leave the cards
before you count the insects. This is not easy
if the insects keep coming and going!

Put your results in a table like this.

Colour of card	Number of insects			
	1	2	3	Average
Red				
Orange				
Yellow				
Green				
Mauve				

Do the test two more times. Move the cards around each time. Write about what you did. Include the results table.

In your conclusion say
a if you think insects like one colour more than the rest,
b what the colours were, starting from the most popular down to the least popular,
c why you think you needed to move the cards each time,
d why walking around them might have helped, and
e why you left the cards in each place for the same length of time.

Extension work

1 The colours on the cards are 'day-glow' bright because they are **fluorescent** (floor-es-sent). Try the experiment again with non-fluorescent cards in the same colours.

2 See if you can use a book to identify the different insects.

3 In your conclusion add the answers to these questions:
a Do the fluorescent colours attract more insects than ordinary colours?
b Are the insects that are attracted to fluorescent colours different from the ones that are attracted to ordinary colours?
c Do different insects prefer different colours?

F S I

Summary

A survey is carried out using different materials that can be felt but not seen.

Equipment

- A feely box
- Some groups may ask for a clock
- A5 graph paper
- A variety of materials can be provided: *cloth* – cotton, fur, chamois leather, nylon, silk, wool; *shapes* – in wood, clay, plastic, etc. (cube, sphere, rectangular block, cylinder, tetrahedron, cone); *metals* – small blocks or squares of sheet iron, copper, aluminium, lead, brass; *non-metals* – cork, slate, perspex, marble, glass, wood. Materials kits, of wood (Philip Harris P10145/5), metal (P10180/7) and solids (P10100/5) contain suitable selections.

Notes

The investigators place six different objects inside the feely box and people can then be tested by putting their hand through the hole to see if they can identify the objects by touch. The feely box is best made out of a strong cardboard box (a suitable size is 40 × 30 × 30 cm) with a 10 cm diameter hole cut in the top. Use strong adhesive tape to secure the box top, otherwise it soon gets torn or comes undone.

During the trials children found it difficult to identify objects unless they were told what the box contained or they were shown pictures of the shapes in the box and asked to pick out particular ones. Children being tested are tempted to peep through the opening unless it is covered with a cloth sleeve or paper ruff.

If an unrelated set of objects is used (like a potato, a pastry cutter, a pumice stone and a travel clock) two difficulties can arise:

1 When one person is tested, others in the room will see what the objects are before they are tested themselves.

2 It is not always possible to find any useful pattern to the results.

When using objects such as different shapes (all wooden, for example), or different materials (all the same shape) it does not matter if the objects have been seen beforehand by the people being tested. The visual recognition of objects only helps with tactile recognition if the objects are distinctly different, such as a wig and an egg.

Children's work

Touch Tests.

Matthew and me made a touch test. We had a cardboard box with a hole in, and we collected Six items. The Six items was a wig, a glue pen, a chess piece, a Shoe horn, a Stick of chalk, army figure legs. Then we asked Ten people from the top class, five people infants classes.

WIG	Glue Pen	chalk	Shoe horn	legs	chess	
✓	✓			✓	✓	MARK
✓						Rachael
✓						Fhraser
✓						EMMA
✓			✓			wayne class1
✓			✓			Sarca
✓			✓	✓		ginca
✓						Lisa
✓	✓		✓			milie
			✓			Richard

most was wig and least was chess piece being picked out.

Report evaluation

Description of investigation: Scores 2 — An attempt is made to explain the method and reference is made to the apparatus used.

Results: Scores 2 — The table shows results that are more or less complete.

Conclusion: Scores 2 — A generally correct conclusion is made.

Total score = 6 marks

How good are people at telling objects apart
just by feeling them – without looking?

Your job is to find out by testing a group of
people.

1 Choose six things and put them in a feely
box.

2 Use a table like this for your results.

Objects					

You can write the names of the objects in
the small boxes at the top of the table.

3 Test people to see if they can tell you what
is in the box – no peeping!

4 Tick a large box in your table if a person
guesses correctly and put a cross if he or
she guesses wrong.

5 Ask at least ten people. Ask 30 if you have
time.

F S I

6 When you have all your results, add up the ticks and write the totals in the boxes at the bottom.

7 Look at the results on display. This shows you how the results can be made into a bar chart.

8 On a piece of graph paper, see if you can draw a bar chart of your results.

9 Write up your experiment using the bar chart to show your results.

In your conclusion say if you think people found some objects or shapes easier to identify than others.

© N.G. Furnell, 1988. Blackwell Education, Oxford

Summary

A standard face is made up with various combinations of eye, eyeshadow and lipstick colours and a survey is carried out.

Equipment

- Face outline sheet
- A4 paper (suitable for water colours if painting)
- Colour pencils, pastels or paints
- Mixing palette ⎫
- Small brush   ⎬ if painting
- Water pot   ⎭
- A5 graph paper

Notes

The factors that make make-up look attractive are investigated here. Six identical 'face outlines' are provided on a master sheet and children can make up the eyes and lips by colouring them in. A survey group of children can be asked to choose the 'made-up face' they like best and the most popular one is identified.

For extension work the most popular made-up face design can serve as a model to produce a new set, where each one is slightly different from the original.

The 'Fair testing' section in the Introduction includes work from this investigation carried out in trials.

Colour pencils give a better effect than felt-tip pens or paints. They can be used to produce the more delicate shades and blends of colour characteristic of make-up.

Water colours are suggested as an alternative medium to use but children may find these difficult for fine details. To keep water colours looking clear and fresh, (a) no more than three colours should be mixed together and (b) brushes should be thoroughly cleaned before changing colours.

The face outlines should be on pink or brown paper rather than white.

Children's work

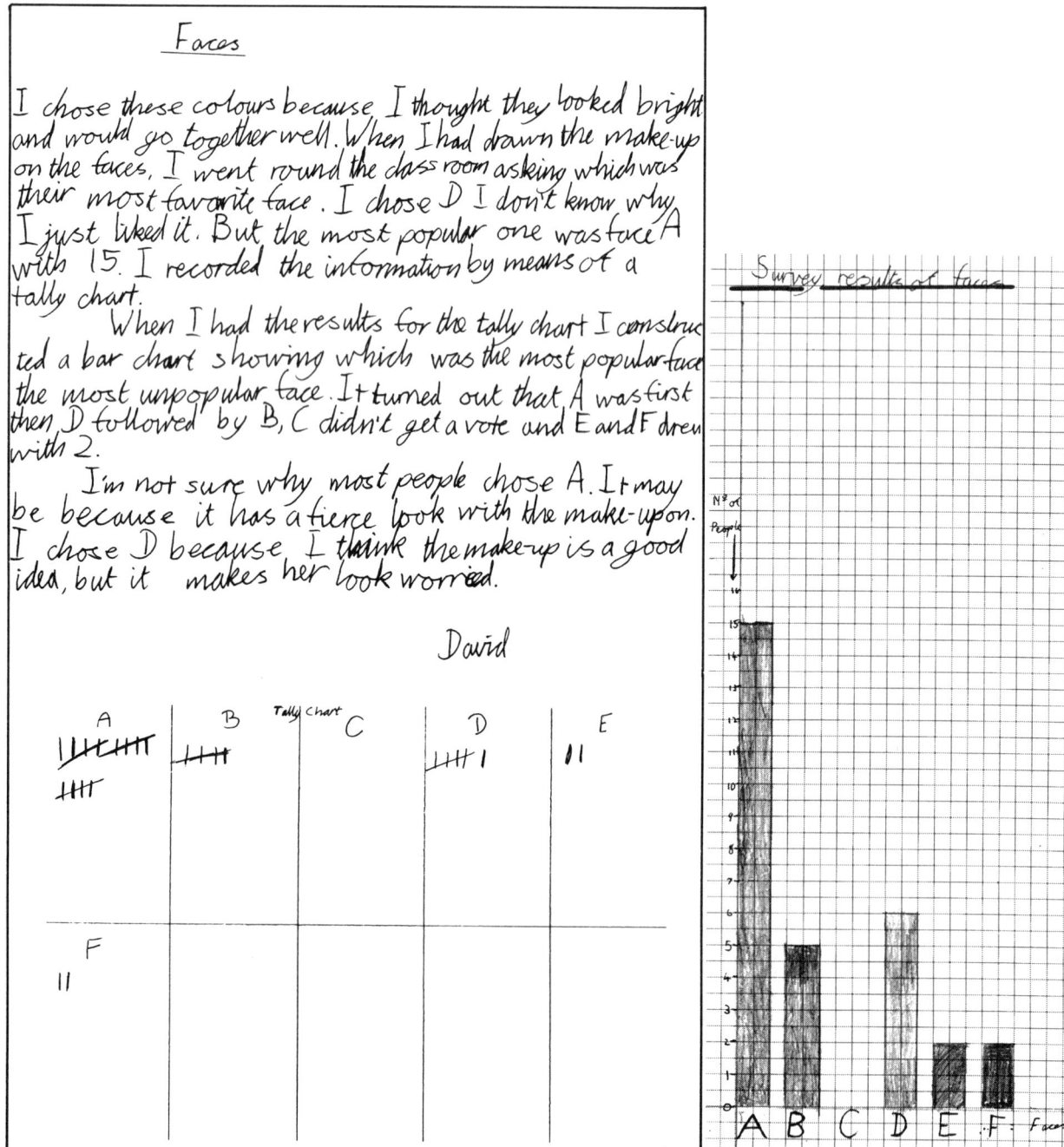

Faces

I chose these colours because I thought they looked bright and would go together well. When I had drawn the make-up on the faces, I went round the class room asking which was their most favorite face. I chose D I don't know why I just liked it. But the most popular one was face A with 15. I recorded the information by means of a tally chart.

When I had the results for the tally chart I constructed a bar chart showing which was the most popular face the most unpopular face. It turned out that A was first then D followed by B, C didn't get a vote and E and F drew with 2.

I'm not sure why most people chose A. It may be because it has a fierce look with the make-up on. I chose D because I think the make-up is a good idea, but it makes her look worried.

David

Report evaluation

Description of investigation: Scores 2 – This contains a generally correct account of method.

Results: Scores 5 – The results are presented graphically. Totals are shown in the bar chart which is properly labelled and the scaling is reasonable.

Conclusion: Scores 3 – What the results show is clearly identified.

Total score = 10 marks

Making-up faces P3

What makes make-up look good?

Your job is to make up five faces and use them for a survey.

You need

- A face outlines sheet
- Colour pencils, felt-tip pens or paints

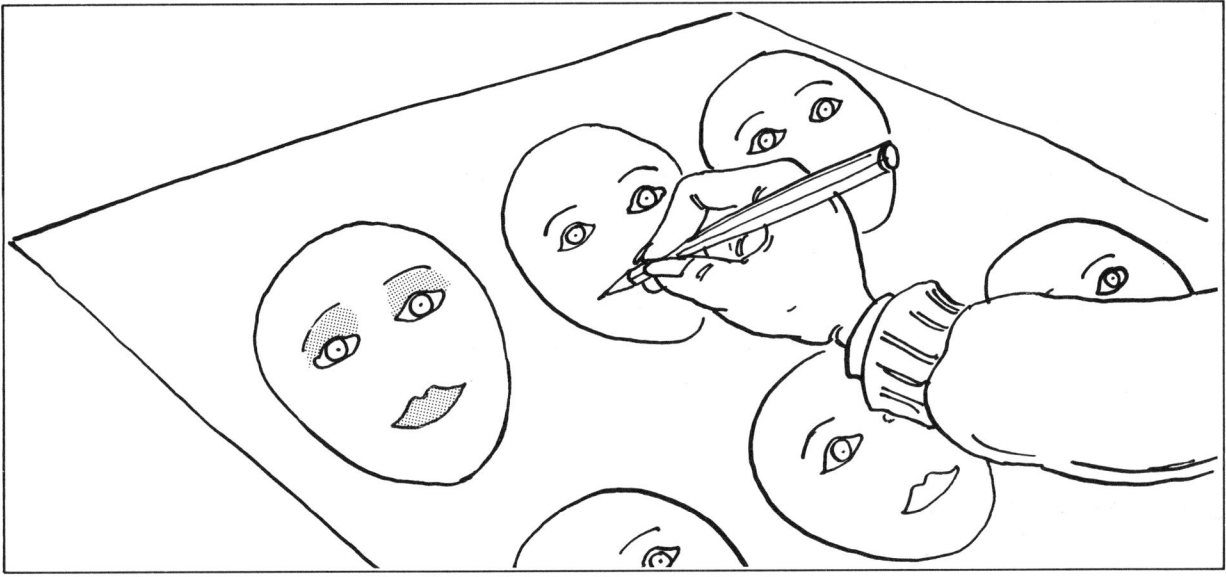

1 Colour each face differently to make up the eyes and lips. Use different colours – from soft pale ones to bright dark ones. The last face is left plain.

2 Label the faces A–E.

3 Have a table like this for your results.

Make-up styles					
Face A	Face B	Face C	Face D	Face E	No make-up

4 Ask a person to choose the make-up they like best.

5 Tick your table for which face they choose.

F **S** **I**

© N.G. Furnell, 1988. Blackwell Education, Oxford

6 Ask 30 people to choose.

7 At the end, add up the ticks and write in the totals.

8 Look at the results on display. This shows you how the results can be made into a bar chart.

9 Ask for a piece of graph paper and see if you can draw a bar chart using your own results.

10 Write up your experiment using the bar chart for your results.

11 In your conclusion say
   a if you think people liked one face more than the rest,
   b how popular the faces are starting from the most down to the least popular, and
   c if people tend to prefer soft pale colours, bright dark ones or both together.

Extension work

Using your most popular face as a plan, make up another set of faces, one the same and the rest slightly different. Get results for these so you have a new 'most popular' face. You could go on to use this as a plan for another set of faces and so on.

In your conclusion say whether your results tell you if make-up is more important for the eyes or for the lips. Explain how you would carry out an investigation to try and find out.

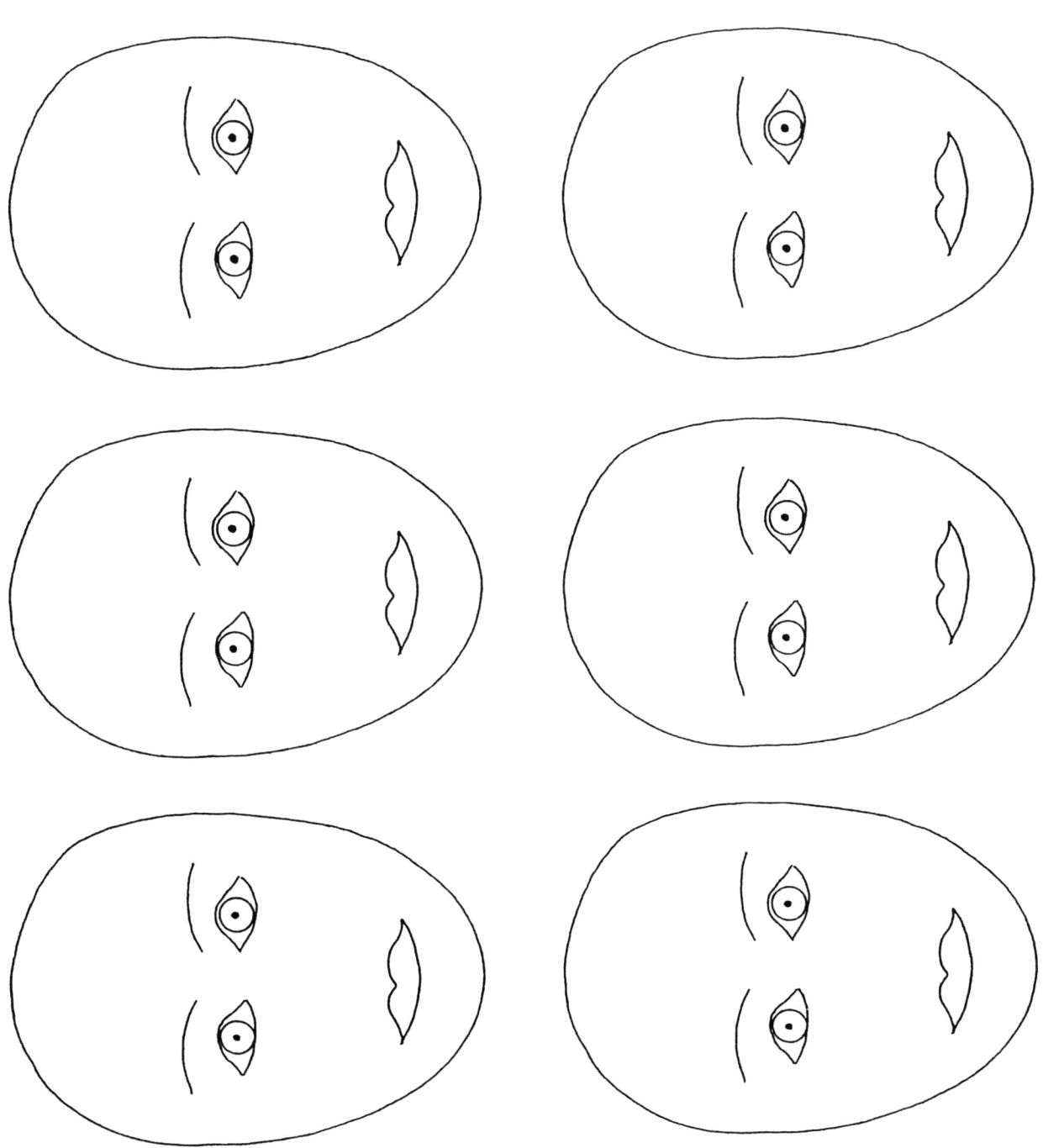

Summary

Different grades of pencil (without markings) are tested by using them for shading. They are then graded according to how dark they shade.

Equipment

- One pencil of each of the following grades: 4H, 2H, HB, 2B and 4B. (The range of grades of pencil can be altered to suit.)
- Optional: spring balance, compression (Philip Harris P11740/2) hardboard mat

Notes

It is recommended that the pencils used are the same brand and from the same batch. The grade markings can be taped over or cut way with a Stanley knife and the symbol (see table) written in its place using a permanent marker pen.

It will help children to start each trial with the pencils sharp and to use the same force to draw with each pencil.

As optional equipment, the hardboard mat on top of the scales provides a writing surface and the means to measure the applied force and keep it constant.

Children's work

<u>H's and B's</u>

I found that by drawing a 5 cm line you could tell which one was which. These are the grades, + = HB, x = 2B, - = 2H, o = 4H and I = 4B. Also one of us (colin) did the test, whilst the other (me) wrote down the scores. We did it this way because if I did one part of the test and Colin did the other part of the test, I might of pressed the pencil down hard and he might of pressed the pencil down lightly and that would make it an unfair test.

Pencil symbol	Score 1–5
+	3
×	4
—	2
O	1
I	5

This is the Pencil Table with the scores from 1-5, with this we learnt which pencil was which.

These are the grades += HB, x=2B, —=2H, o= 4H, and I =4B. We checked on the chart to make sure that all of these were right. After we had done the test we made sure by doing another test.

Extension Work

Pencil symbol	Shading score 1–5				
	Trial 1	Trial 2	Trial 3	Trial 4	Average
+	3	3	3	3	3
×	4	4	4	4	4
—	2	2	2	2	2
O	1	1	1	1	1
I	5	5	5	5	5

I think that out of all the pencils we have tested 4B would be best for sketching. I also think that the 2H pencil is good for technical drawing. HB is good for writing.

When we did the test we did four tests, because if we did one it might have been wrong, but if we did four test and the results were all the same, then we must have been correct. We also found out which pencil is which, the += HB, x=2B, —=2H, o =4H, I=4B and by doing these four tests, we found out that these were all right.

Report evaluation for investigation

Description of investigation: Scores 2 – An attempt is made to explain the method and there is some indication of the purpose of the investigation.

Results: Scores 2 – The table of results is complete.

Conclusion: Scores 3 – What the results show is clearly identified.

Refinements: Scores 2 – Precautions are taken to improve accuracy (only one person tested the pencils) and further action is taken on completing the investigation (the results were checked by carrying out another test).

Total score = 9 marks

Report evaluation for extension work

Results: Scores 3 – This consists of a table showing averages.

Conclusion: Scores 4 – This has an interpretation based on results.

Refinements: Scores 1 – This shows understanding of the limitations of results (the reason given for doing four tests).

Total score = 8 marks

The score for 'Description of investigation' might also be included here since the same procedure is used in each case.

Pencils are made in different grades for different jobs. Your job is to test a set of pencils and choose the grades you would use for doing different things.

Unfortunately the pencils you have for this work have lost their labels!

You will have to do a 'pencil test' trial to grade them.

1 Score the pencils from 1 to 5 for how dark they draw with 1 for the lightest and 5 for the darkest.

2 Use a table like this for your results.

Pencil symbol	Score 1–5
+	
×	
–	
o	
I	

3 Write about how you did your experiment and use the table to show your results.

4 The pencils you have are grades 4H, 2H, HB, 2B, and 4B. 4H gives the lightest shade, then 2H, then HB, and then 2B. The last one, 4B, gives the darkest shade. From your test, can you say which grade each pencil is?

5 In your conclusion, write down which grade each pencil is.

Extension work

Use the same method to get four sets of results.

Put them in a table like this.

Pencil symbol	Shading score 1–5				
	Trial 1	Trial 2	Trial 3	Trial 4	Average
+					
×					
—					
O					
I					

Write up your results.

Say in your conclusion:
a which grade pencil you would use for sketching,
b which grade you would use for technical drawing,
c which grade you would use for writing, and
d why you think four sets of results are better than just one set.

This experiment gives B pencils the highest scores. Say why you think this is unfair and how you would make the experiment fairer.

© N.G. Furnell, 1988. Blackwell Education, Oxford

Summary

Common scented liquids are used in a survey to see if some scents are more easily identified than others.

Equipment

- Safety glasses (see below)
- 20 cm^3 bottles of: perfume (block form may be used if preferred), aftershave (or roll-on deodorant), peppermint essence, lemon essence, almond essence
- Blotting paper or paper towel
- Six dishes (tin or polythene lids)
- Tray
- Scissors

For extension work
- 10 cm^3 measuring cylinder (Philip Harris C31760/6)
- Six small containers or 100 cm^3 beakers (Philip Harris C17500/9)
- 3 cm^3 graduated dropping pipette (Philip Harris C61502/2)
- Marker pen for labelling containers

Warning The essences suggested are commonly available and sold for cake making. However, some important points concerning safety should be remembered:

1 Safety glasses should be worn by anyone in the vicinity when liquid essences are being poured.

2 Essences, perfumes and aftershaves in liquid form may contain volatile solvents. Do not use them in the vicinity of burners, cookers or heaters.

3 Ensure that substances in liquid form which could be a source of solvent abuse are stored securely. If preferred, restrict the range of substances to solid forms like bath salts.

Notes

This is a test of our ability to identify the smells of common substances like perfume, aftershave and fruit essences on squares of paper so that there is no visual clue to their origin.

Squares of blotting paper (or paper towel) can be cut to fit the dishes and labelled underneath in pencil with the name of the scent to be added. An unscented square is included to see if there are some people who cannot distinguish the smell of the paper from the scents being tested. Children can then be tested by asking them to smell each square in turn.

In trials we found that children were unable to say what even quite familiar smells were unless we told them what substances were being used.

The smells tend to intermingle if the squares are handled so it is best to keep them in the dishes. Tin lids are ideal. Initially we used disposable Petri dish lids but the liquid essences dissolved the plastic!

Avoid giving children full bottles of liquids which may get dropped or spilled. Ideally keep duplicates and 1/4-fill the experimental bottles from the stock bottles as required.

Wear safety glasses when pouring the liquids.

For groups wishing to carry out part 2 of the extension work, the following are suggestions for materials:

Odour	Material
Burnt	Burnt wool or feathers
Floral	Scented flowers e.g. rose (or lavender water)
Foul	Stale milk (very small quantity)
Fruity	Pear drops (or fruit essence)
Resinous	Cedarwood (a wood offcut or oil extract)
Spicy	Curry powder

Children's work

The objectives of this experiment are to distinguish how well thirty children from the sixth year department could identify six different smells. We set about this, first by making a chart for our results. We then decided not to tell the children which scents were being used. Thirdly we set out the saucers on a tray. We used blotting paper to soak up the liquids and hold the smell. We then asked ten children from each of the three classes to try to identify the smells. The scent most identified was the unscented blotting paper as you can see from the graph. Twenty people identified this. Going down from the highest number of correct answers to the lowest would be, unscented, peppermint, lemon, perfume, aftershave and almond. We found that a lot of people identified 'almond' as 'marcipan'. Only one person guessed correctly. A lot of people thought the aftershave smelt like soap. Personally I agree with them. I would have thought the lemon would have been identified by all of them because lemon has such a distinctive smell. About ten people thought the perfume was Aftershave. Nineteen people identified the peppermint. Nineteen was the second highest number of correct answers. We thought a lot of children would be able to identify this because they are likely to have had peppermint sweets at some time in their life, also they are likely to have used peppermint toothpaste. A few people guessed toothpaste as their answer for this scent.

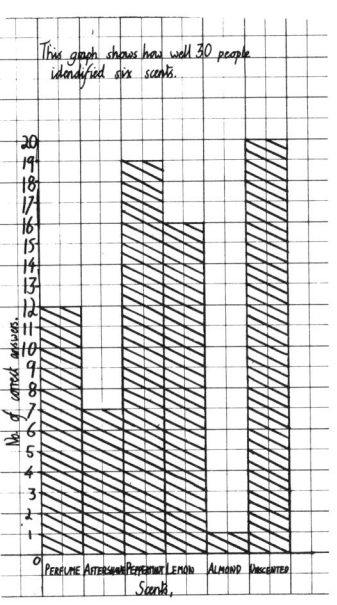

Evaluation of investigation

Description of investigation: Scores 5 – The purpose of the investigation is correctly stated, the apparatus used is described and the method is clearly expressed.

Results: Scores 5 – The table of results with totals and good graphical presentation.

Conclusion: Scores 4 – The interpretations (about peppermint) are based on the results obtained. The results are consistent with the investigators' predictions and have been interpreted accordingly.

Total score = 14 marks

To start our extension work we used the dropping pipette to measure percentages of perfume into a ten cms³ measuring cylinder. The percentages were 0% perfume, 0·01% perfume, 0·1% perfume, 1% perfume, 5% perfume and 10% perfume. We asked thirty children to tell us whether they could smell the perfume in the liquid. As you can see from the graph, only one person did not guess correctly the 10% solution, 28 people guessed correctly the 5% solution, 24 people guessed correctly the 1% and the 0·1% solutions, 27 people guessed correctly the 0·01% solution and 10 people guessed correctly the 0% solution. I think that the 20 people who did not guess the 0% solution could probably smell the other solutions next to it. Some other people had bad colds which could have affected their sense of smell.

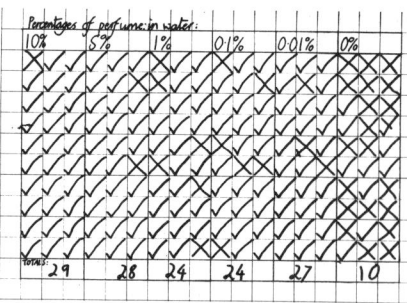

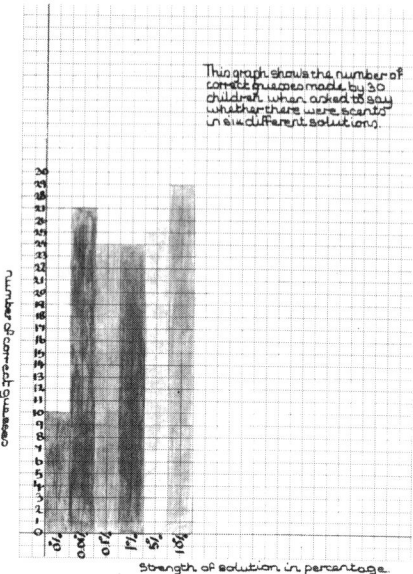

This graph shows the number of correct guesses made by 30 children when asked to say whether there were scents in six different solutions.

Evaluation of extension work

Description of investigation: Scores 3 — A correct account of method is given and reference is made to the apparatus used.

Results: Scores 5 — A table of results with totals and good graphical presentation.

Conclusion: Scores 4 — The interpretation (of why so many people did not identify the 0% solution) is based on the results obtained.

Only 10 people out of 30 correctly identified the unscented solution. This means that two-thirds of the people said they could smell the scent even when it was absent. For solutions containing the scent, results would have to be above this two-thirds level to show the scent *was* being detected. The two-thirds level may vary from survey to survey but the principle remains the same.

Refinements: Scores 1 — An understanding of the limitations of the results is shown (e.g. the idea of colds affecting the sense of smell).

Total score = 13 marks

Common scents

Can people tell substances by their smell?

Your job here is to see how well people can identify some common substances using their sense of smell.

Wear safety glasses when pouring the liquids

1 Use a table like this for your results.

Scents					
Perfume	**Aftershave**	**Peppermint**	**Lemon**	**Almond**	**Unscented**

2 Before you start, decide if you are going to tell people which scents are being used, or just let them try to guess.

3 Ask about 30 people to try and identify the different scents.

4 In the large boxes in your table, put a tick for a correct guess and a cross for each one wrong.

5 At the end add up the ticks and write the totals in the boxes at the bottom.

F S I

6 Look at the graphs on display. These show you how the results can be made into a bar chart.

7 Ask for a piece of graph paper and see if you can draw a bar chart using your results.

8 Write up your experiment using the bar chart to show your results.

9 In your conclusion:
a list the scents – from the one identified the most to the one identified the least and
b say why you think people found some scents easier to identify than others.

Extension work

1 How sensitive is our sense of smell? You could test this in the following way using an essence that dissolves in water.
a Use a 3 cm³ dropping pipette and a measuring cylinder to make a 10% solution (1 cm³ of essence and 9 cm³ of water).
b You could then make a 1% solution (1 cm³ of the 10% solution and 9 cm³ of water) and so on.
c Make up a set of solutions of different strengths (such as 0.01%, 0.1%, 1%, 5%, 10%) and one of plain water.
d Carry out a survey by asking the question, 'Is the scent present or absent in each liquid?'
e Record the number of correct answers for each strength solution.
f Use your results to plot a graph of **Strength of solution (%)** (along the *y*-axis) and **Number of correct answers** (along the *x*-axis).

2 Do you think that smells liked by some people are disliked by others?

© N.G. Furnell, 1988. Blackwell Education, Oxford

Some scientists say that there are six different types of smell: burnt, floral, foul, fruity, resinous and spicy.

See if you can think of an experiment to find the order for the most popular to the least popular of these different types of smell.

Righting wrongs

Summary

Different methods of correcting handwriting and drawing mistakes are tested and the best buy determined.

Equipment

- Ink rubber: Any coarse-type ink eraser
- Typists' rubber: The pencil type or flat hexagonal shape are both suitable
- Natural rubber (as opposed to the plastic type)
- Tipp-Ex paper: White correcting paper (sold as a typing accessory) is available as a roll or as a packet of strips. It is quite expensive so only a small piece should be provided
- Tipp-Ex pen: White correcting fluid sold in bottles often contains solvents which make it unsuitable for children. Correcting pens with sealed reservoirs are safer and more convenient for children to use. Tipp-Ex waterbase fluid is solvent-free and could be used as an alternative to a correcting pen
- Eraser pen: As sold by Helix Ltd
- Paper
- Pencil
- Ball pen
- Fountain pen
- Felt-tip pen

Washable cartridge pen blue ink and erasable fibre-tip ink is removed by the eraser pen so at least one of these should be used. It is quite acceptable that children use their own pens and pencils instead of the ones you provide.

Procedure

Any suitable method is acceptable. For example, squares (roughly 1 × 1 cm) can be drawn onto the paper and filled in in pencil, ball pen, and fountain pen. A test to remove them can be carried out. Use a fresh set of squares for each correction method.

Children's work

Rubber Tests.

Method.
I was setting an experiment for testing different rubbers. I had to use an ink rubber, typist rubber, natural rubber, tipp-ex paper and a tipp-ex pen to rub out pencil, ball-point pen, felt-tip pen and a ink pen.

Results
I found that the best rubber was the tipp-ex pen. Although this pen left a small bump it didn't leave any of the pen/pencil marks.

Conclusion.
I think that the Ink rubber and the natural rubber should be improved because they both leave marks on the paper and they don't rub out on some pens/pencils.

Correction method	Score out of 10				Average score
	Pencil	Ball pen	Felt tip	Ink	
Ink rubber	0	3	7	5	3·75
Typists' rubber	7	9	8	10	8·5
Natural rubber	0	3	6	0	2·25
Tipp-Ex paper	10	6	9	10	8·75
Tipp-ex pen	10	9	10	10	9·75
Eraser pen					

Report evaluation

Description of investigation: Scores 2 – An attempt is made to explain the method and reference is made to apparatus used.

Results: Scores 3 – This consists of a table showing averages.

Conclusion: Scores 2 – A generally correct conclusion is made. (The comment made in the conclusion is an opinion based on the results.)

Total score = 7 marks

We all make mistakes – but which is the best
way to correct a written mistake?

In the table below, there are six ways we
could try.

See if you can make up an experiment to test
them out.

Test each method. Make sure it is a fair test.

Give each method a score out of 10 for each
test.

Put your results in a table.

See if you can find the
averages using a calculator.

Write up your experiment.

In your conclusion see if you
can suggest some ways to
improve the experiment.

Correction method	Score out of 10				Average score
	Pencil	Ball pen	Felt tip	Ink	
Ink rubber					
Typists' rubber					
Natural rubber					
Tipp-Ex paper					
Tipp-ex pen					
Eraser pen					

FSI

Summary

Sweets are used in a survey to see if some colours are more popular than others. The position of chosen sweets is recorded to see if people tend to choose sweets because of their position in a line.

Equipment

- Coloured sweets: red, orange, yellow, green and purple. Twenty of each colour should be ample (only 30 are eaten in the survey)
- A sweet 'tray': a strip of card with five places for sweets labelled with numbers from 1 to 5. A strip from an egg tray (keyes tray) can be used for sweets that roll
- A sweet 'mixing bag': a small clean bag that is not transparent. A cloth bag with a drawstring is ideal
- A5 graph paper

Notes

Sweet manufacturers want sweets to look as appealing as possible and their colour is an important factor. Here sweets are used in a survey to see if some colours are more popular than others. The position of chosen sweets is also investigated to see if a person's choice is influenced by a sweet's place in the row. The sweets are put in the row in random order and the children being tested pick one each. The sweets are 'randomised' before each person is asked. A way of doing this is given in the pupils' page.

Wrapped sweets are more hygienic to use than unwrapped ones. The cheapest wrapped sweets are sold in packets of various flavours and colours – for example, Chewits, manufactured by Famous Names Ltd. (Flavours are packed individually in appropriately coloured wrappers.)

Unwrapped sweets may be cheaper but have to be purchased in assorted colours. Possible brands to use are Tom Thumb Drops or Chocolate Beans (both manufactured by Bonds of London).

Children's report of the extension work

We were asked the question "How many people should be asked in a real survey?"

I think one hundred people should be asked in a real survey.

By conducting a survey in the following way we can try to answer the question.

We had to use two colours of sweets, the one we found to be most popular and the one we found to be least popular. As we had two most popular sweets in our first survey we had to use both of them as well as the least popular sweet. The two most popular sweets were purple and red. The least popular sweet was yellow.

We decided to ask seventy people to choose a sweet from the selection of three. I dropped the sweets down rather than having them in a row because if they were in a row the person choosing might pick a sweet because of the position it was in and in this test position is not important. Once the person had chosen we ticked a chart of the colour of sweet chosen

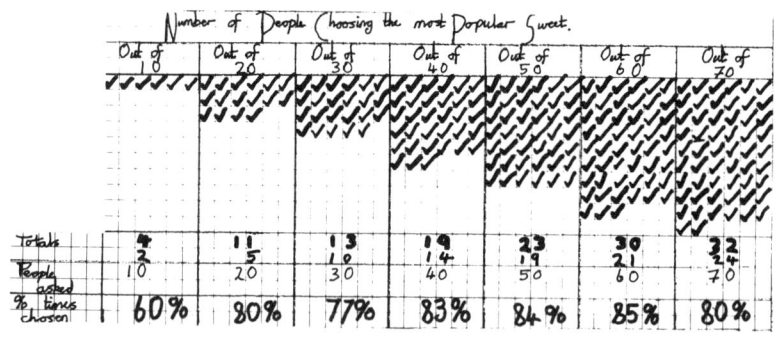

Number of People Choosing the most Popular Sweet.

	Out of 10	Out of 20	Out of 30	Out of 40	Out of 50	Out of 60	Out of 70
Totals	6	11	13	19	23	30	32
People asked	10	20	30	40	50	60	70
% times chosen	60%	80%	77%	83%	84%	85%	80%

Each time someone chose one of the two most popular sweets we ticked our chart. If they chose yellow I kept a record of it. Once we had finished asking the ten people we did not tick OUT OF TEN anymore.

We had another chart to keep a record of how many people had been so as not to get confused. Once we had finished our survey we worked out the percentages of each stage. Our graphs shows clearly the percentage number of times the most popular sweet was chosen by a given number of people.

Our chart evens out at 40 and 50 people.

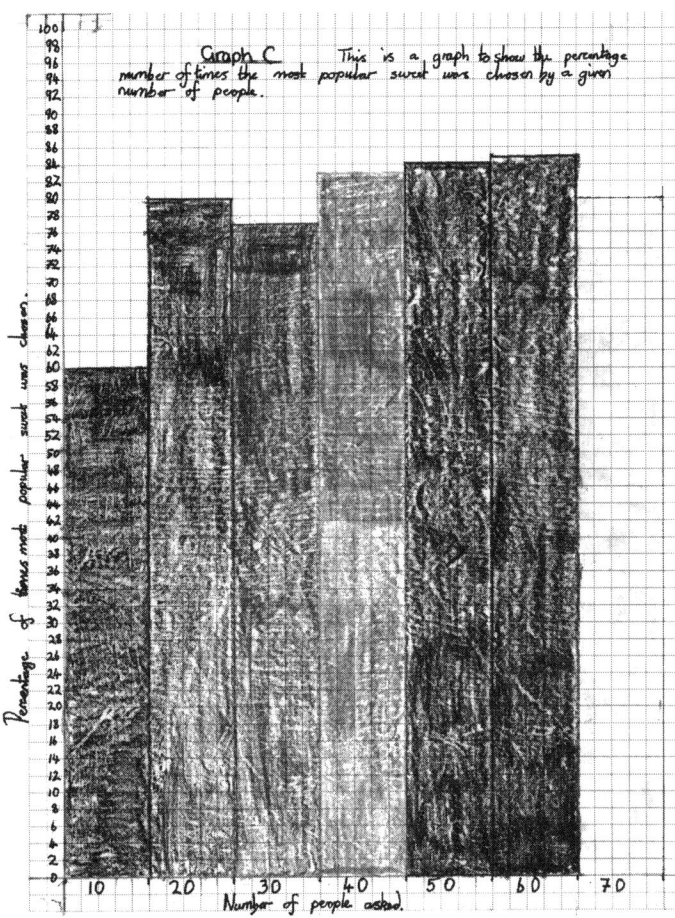

Evaluation of extension work

Description of investigation: Scores 5 – The purpose of the investigation is correctly stated, the apparatus used is described, and the method is clearly expressed.

Results: Scores 5 – This consists of a table with totals and good graphical presentation.

Conclusion: Scores 4 – The interpretation (of where the bar chart 'evens out') is based on the results obtained.

If two colours that were equally popular were used, each would have a 50% chance of being chosen. If the three colours of sweet used here were equally popular, each would have a 33% chance of being chosen. The chance of any two of them being chosen would be 66%. The bar chart 'evens out' at around 82% indicating a strong preference for choosing the two most popular colours.

Refinements: Scores 1 – Overcoming a minor difficulty is explained (having a tie for the most popular colour of sweet).

Total score = 15 marks

Which colour sweet do you pick from a bag of coloured sweets? Is your favourite colour sweet tops with everyone?

If people are offered a sweet from a row of sweets does the middle one get picked more often than the end ones?

Your job is to find out the answers by carrying out a survey.

You need

- Sweets of five assorted colours
- A sweet tray
- A 'mixing-up' bag

1 Have tables like this for your results.

Colour of sweet chosen				
Red	Orange	Yellow	Green	Purple

Position on tray of sweet chosen				
1	2	3	4	5

Have very clean hands, please!

FSI

2 To make it a **fair test** the sweets have to be mixed up before they are put on the tray.

    a Put five sweets, each one a different colour, in a bag. Close the bag and mix the sweets up.

    b Without looking, take one sweet out at a time.

    c Put them on the tray in the order that you take them from the bag.

3 Ask someone to choose one sweet from the tray.

4 In your tables, tick the colour of the sweet chosen and its position on the tray.

5 Ask about 30 people to choose sweets. (Repeat steps 2–4 each time.)

6 At the end add up the ticks and write the totals in the boxes at the bottom.

7 Check the bar charts on display.

8 Ask for a piece of graph paper and make up bar charts using your results.

9 Write up your experiment using the bar charts to show your results.

10 In your conclusion, say:

    a if you think people like one colour more than the rest

    b what the colours are in order of popularity

    c if people tend to choose sweets from the same place in the row.

Extension work

How many people do you think should be asked in a real survey?

F **S** **I**

Here is one way that we can try to answer this question.

1 Take the sweet that was most popular and the one that was least popular.

2 Ask people to choose one of the two sweets.

3 Draw up a results table like this. (It will have to be larger if you ask more than 60 people.)

	Number of people choosing the most popular sweet					
	Out of 10	Out of 20	Out of 30	Out of 40	Out of 50	Out of 60
Totals						
People asked	10	20	30	40	50	60
Times chosen (%)						

a For the first 10 people, each time the most popular sweet is chosen, tick each big box.

b For the next 10 people, tick all the boxes except the first one, and so on.

4 You will also need a table to record the number of people that choose the least popular sweet.

5 After asking 60–100 people add up the ticks and write the totals in the second row of boxes.

6 Use your results to plot a graph of **Number of times the most popular sweet is chosen (%)** (along the *y*-axis) and **Number of people asked** (along the *x*-axis).

7 The region where the graph becomes level is the answer to the question, 'How many people do you think should be asked in a real survey?'

There is no one answer – it varies for different surveys!

F S I

© N.G. Furnell, 1988. Blackwell Education, Oxford

Bird feed

Summary

Bread cubes are dyed with different food colours and put out for birds to eat.

Equipment

- Three slices of white bread (medium)
- Knife for cutting the bread into cubes
- Liquid food colouring: red, orange, yellow, green, blue
- Teaspoon
- Six dishes or bowls
- Tray

Notes

Birds are often seen to descend onto school playgrounds after playtimes to finish off the crisps and sandwiches!

To see if birds show any colour preference in what they eat, bread cubes are dyed with different food colours and put out for the birds to feed on. If the cubes are scattered in an open area such as a grassy verge or the playground, birds will soon be attracted to them.

Some artifical food colours (like yellow and green) may contain additives such as tartrazine (E102). Natural food colours are reasonably priced alternatives that are sold in health food shops (or contact Rayner-Burgess Ltd, 4 Bull Lane, London N18 1TQ).

It is advisable to begin this investigation early in the day. The first time we tried it, the birds took 3 hours to eat about half the cubes. They were suspicious of technicolour bread and would not eat any of the blue or green cubes. However, they do get used to it. In very cold weather when food is scarce, the cubes are accepted far more readily than in the summer.

The cubes need to be put in a place where (a) birds feed undisturbed and (b) uneaten cubes are easily seen. Choose a site where counting up uneaten cubes will be easy – they can become hidden in weeds or long grass.

Ensure that children work safely – away from traffic and where they can be seen and supervised.

T8

This investigation should not really be carried out during the summer term when birds have young. Dry bread is a potential hazard to fledglings whose only water supply is from moist foods like grubs and worms. Furthermore, bread is unlikely to provide the balance of nutrients found in their natural food sources.

MAKING BREAD CUBES

If the cubes are cut small (64 from a 'medium' slice) they are more likely to be eaten completely. (When partly eaten cubes are left it makes counting difficult.) A suitable mix of bread and dye is ½ slice of bread (30 cubes), four teaspoons of water and ½ teaspoon colouring. Birds will feed on wet or dry cubes. In very cold weather wet cubes freeze!

Teacher's Notes

Bird feed

You are going to give a party for the birds!

Your job is to find out what colour treats the birds like best.

1 Make some cubes from bread. **Mind your fingers.**

2 Use the food dye to colour them.

3 Ask permission to go out and put the food out for the birds.

4 Use a table like this to record your results.

Colour	Number of bread cubes		
	at start	at end	eaten
Red			
Orange			
Yellow			
Green			
Blue			
Plain			

Did you have the same number of cubes for each colour?

Did you scatter the cubes on the grass so that all the colours were mixed up?

Can you do a better experiment to check the results of this one?

F **S** **I**

5 Write about your experiment using the table for your results.

Extension work

Repeat the experiment each day for 5 days.

In your conclusion write about any changes in your results.

F S I

Summary

A small mammal is given a selection of different foods to eat for a set time. The mass of each food eaten is determined.

Equipment

- Small mammal(s), preferably two or three (all of the same species). Hamsters, rats or guinea pigs could be used
- Balance (to 250 g)
- Clock
- Scoop or spoon
- Six-section paint palettes (one per animal)
- Plastic tanks or similar (with water bottles) to put each animal with its food
- Some form of lid (e.g. wire mesh) may be required to stop animals from climbing out
- Six types of food. The following foods are convenient as they 'keep': Cat stars (fish meal), dried peas, flaked maize, oats, peanuts, sunflower seed, wheat. Some of these may not be suitable for the animals used; if so, other foods (fresh or dried) should be used accordingly.

Notes

This is a firm favourite with children who enjoy working with animals. The investigation can be carried out in a variety of ways. One method is outlined below.

Each food is weighed out in 10 g amounts and put into the palette sections. One palette is needed per animal, and each palette should contain the same food. The animals can be left to feed in the tanks for about 30 minutes. After this period they are returned to their cages and the uneaten food weighed. Food that was scattered, including empty shells, should be included in this weighing, but it may be soiled and should not be put back in with uneaten food. Utensils and feeding tanks should be washed out after use.

Prior to an investigation period animals should have a staple food (such as grass pellets, in the case of a guinea pig) in their cage. They are then more likely to feed from the variety of foods offered in the test.

It is not recommended that animals be kept specifically for these activities. Animals could be borrowed from another school on a short-term basis so that children could experience working with them.

A stable container must be used, otherwise food may be tipped out as animals move around.

The balance needs to weigh to the nearest gram. An electronic top pan balance is ideal.

If animals have individual cages feeding tanks will not be needed.

Pairs or groups of gerbils live together quite happily in the same cage provided they are put together as juveniles. However, if they are separated, even for as short a time as 20 minutes, they can become aggressive towards each other and bite the children handling them. This is likely to happen if they are taken from the cages separately by children and effectively 'isolated'.

References

Small mammals in schools, December 1986, RSPCA
Be safe, 1986, Association for Science Education

Your job is to find the favourite food of a small mammal.

Feed two or three of the same mammal if you can and get an average result.

Never put your fingers through the bars of a cage.

Do not make sudden movements.

Handle animals very gently .

Use a table like this for your results. You will need a table for each animal.

Type of food	Mass of food (grams)		Amount eaten (grams)
	before feeding	after feeding	

If you feed more than one animal put the results together at the end in a table like this.

Type of food	Amounts eaten (grams)			
	Pet 1	Pet 2	Pet 3	Average

See if you can find the averages using a calculator.

Leave the equipment clean and tidy.

Always wash your hands after handling animals.

Write up your experiment fully using the table to show your results. If you used two tables in your experiment, you only need to present the second table.

In your conclusion list the foods from the most popular to the least popular.

F S I

Summary

Bubbles are blown and their diameters measured for different strengths of detergent solution.

Equipment

- Detergent solutions in water of the following strengths (see notes): 5%, 10%, 15%, 20%, 30%, 40%, 50%, 60%, 70%, 80%, 90%, 100%
- Bubble blower (as supplied with 'Bubble Liquid' in toy shops)
- Ruler
- A5 graph paper
- Large plastic tray or sink
- Wiping-up cloth

For extension work
- 10 ml measuring cylinder (Philip Harris C31760/6)
- 100 ml measuring cylinder (Philip Harris C31820/9)
- 100 ml beaker or plastic pot (Philip Harris C17500/9)
- Detergent (any household washing-up liquid)
- Water supply

Notes

Bubble blowing, the basis of this activity, makes it a very popular investigation. Children blow bubbles using the detergent solution provided in different concentrations. From each solution, five bubbles are blown and their diameters are measured with a ruler as soon as they leave the bubble blower. The average diameters for each solution are then calculated.

For extension work, new solutions are made up and tested as the experiment progresses.

A large tray or sink for working in and a clean rinsed cloth for wiping up are important. Unless carefully supervised, children can quickly make their workplace messy and the floor slippery.

Prepared solutions are best made up in plastic pots with secure lids.

The amount of detergent you use gives the percentage strength if water is added to make exactly 100 cm^3 of solution.

Examples: for a 5% solution, use 5 cm^3 of detergent + 95 cm^3 of water; for a 20% solution, use 20 cm^3 of detergent + 80 cm^3 of water.

Your job is to discover how to make a really good bubble mixture.

1 You need a table like this for your results.

Detergent solution strength (%)	Bubble diameter (cm)				
	1	2	3	4	Average
5					

2 Try out the 5% solution first. Test the mixture by blowing four bubbles. **Blow your bubbles over a tray or sink. Do not blow bubbles over the floor.**

3 Write your answers in your table along the first row.

4 See if you can calculate the average bubble size. Write your answer in the table.

5 Choose another strength of solution and blow another set of bubbles. Put the results in your table.

6 Find the new average size and write it in your table.

F **S** **I**

            © N.G. Furnell, 1988. Blackwell Education, Oxford

7 Use four other strengths to get six sets of results.

8 Write up your experiment using the table to show your results. In your conclusion say what strength of detergent solution made the best bubbles.

Extension work

To find the ideal strength solution for bubble blowing you will have to make up solutions yourself.

1 Rinse and dry your hands if they are soapy.

2 Ask for a piece of graph paper and label it like the graph on display.

3 Mark the average bubble diameters onto your graph from your first experiment.

4 Make up a few of your own solutions to any strengths you want to try out.

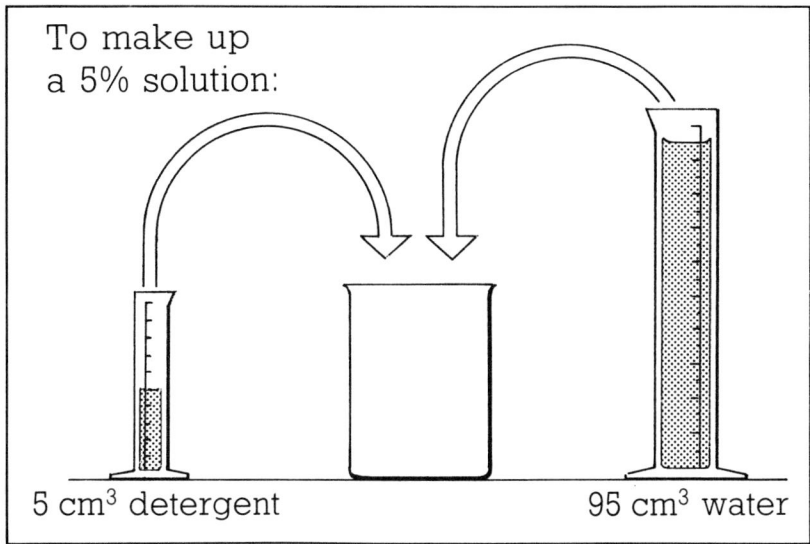

To make up a 5% solution:

5 cm³ detergent   95 cm³ water

To help mix it, pour it back and forth into the measuring cylinder (do not shake).

To make up a 12% solution, use 12 cm³ of detergent + 88 cm³ of water.

The amount of detergent you use gives the percentage strength if water is added to make exactly 100 cm³ of solution.

F S I

5 Test out your solutions.

6 Add the new results to your graph.

7 Write up your results. In your conclusion, state:
    a the ideal strength of detergent solution for bubble blowing and
    b some things you could change in the experiment to get more accurate results.

F S I

Summary

Cycle brake blocks for aluminium and steel wheels
are tested on aluminium and steel surfaces in dry and
wet conditions. Average pulling forces are calculated.

Equipment

- Pairs of brake blocks for aluminium wheels (e.g.
  Weinmann) and steel wheels (e.g. Raleigh Rain
  Check (leather/rubber combination))
- Two switch-backing boxes – double gang metal
  (13 cm × 7 cm × 17 mm or nearest equivalent)
- 1–5 kg mass
- Clean, grease-free trays or sheets (around
  25 × 40 cm) of aluminium (2–3 mm thick) and plated
  steel (1–2 mm thick)
- Spring balance, Newton scale (Philip Harris
  P11640/9 is ideal for the 5 kg mass but P11460/7 and
  P11480/2 together will do. With a 1 kg mass
  P11440/1 is suitable.)
- A5 graph paper
- Wiping-up cloth

Notes

Cycle wheels are made from materials such as
aluminium, plastic and chromium-plated steel. Each
kind of wheel has its own types of brake block and
this investigation compares two of them acting on
different surfaces.

The brake blocks are fitted to a 'brake block box' as
described below. This can be placed at the end of one
of the metal sheets and pulled along steadily with the
forcemeter to obtain a reading. Some children find that
the box tends to judder or topple when they first start
to pull it. They generally get the hang of it once they
have a few 'test pulls'. To test the blocks on a wet
surface about 25 cm^3 of water is poured onto the plate.
Children need to make sure that the blocks stay
moving on the wet surface.

The differences in pulling forces are largest with the
5 kg mass but a smaller mass may be more convenient
to use and safer for small children.

The braking effect of leather-faced blocks is good in
wet weather whereas rubber blocks lose their
efficiency. Children should see this from the
experimental results. Their results may also suggest
that leather-faced blocks are preferable for aluminium

wheels. They are in fact unsuitable because deposits from the aluminium clog the leather. You may have to scrub them periodically with soapy water to remove this. Alternatively, keep a set of leather-faced blocks for the steel sheet and another for the aluminium sheet.

On page 66 children are asked to suggest a better way to test out the brake blocks. It can be explained to them that 'normal use' is a much better way to test the brake blocks (fitted to the type of cycle they are intended for) because of the difficulty of accurately reproducing real conditions.

THE BRAKE BLOCK BOX

Switch-backing boxes serve as sturdy but inexpensive mountings for the brake blocks. They are sold in electrical and hardware shops. The holes ready-drilled in the base will need enlarging slightly (to around 6 mm) for the brake shoe bolts.

Two brake blocks fitted to the central holes each side of the back make the box reasonably stable even with a 5 kg mass in place.

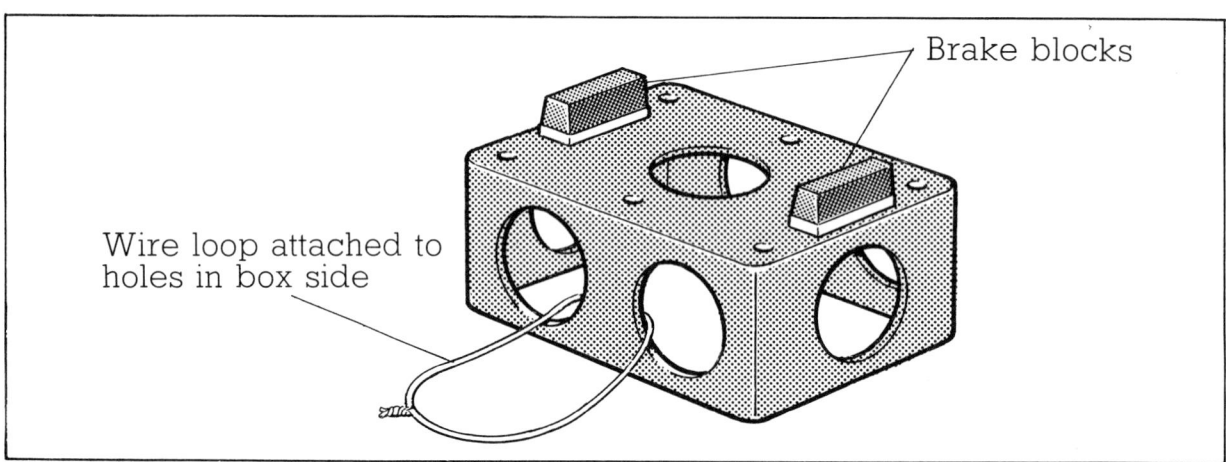

Brake blocks

Wire loop attached to holes in box side

If desired, for greater stability, four brake blocks could be fitted into corner holes.

Steel wheels are chromium-plated – that is, the braking surface is chromium, not steel – and this should be provided for in this experiment. It is a simple and fairly inexpensive process to have steel plated in decorative chrome and polished. (Look up 'Electroplaters and metal finishers' in your Yellow Pages directory.)

If ordinary mild steel sheet is used it will soon go rusty. Stainless or tin-plated steel are not really suitable because they do not have a covering layer of chromium.

Do not experiment with brake blocks on a cycle – Ask for advice at your local cycle shop

On your bike your brakes can make the difference between stopping safely and an unplanned trip to Casualty. When it's wet, some brakes, or rather brake blocks, are useless.

Your job is to test out two kinds of brake block to find the best sort to use.

You need

- Newton forcemeter
- Trays or sheets of aluminium and chrome-plated steel
- Brake blocks on brake block boxes
- Water
- 1–5 kg mass

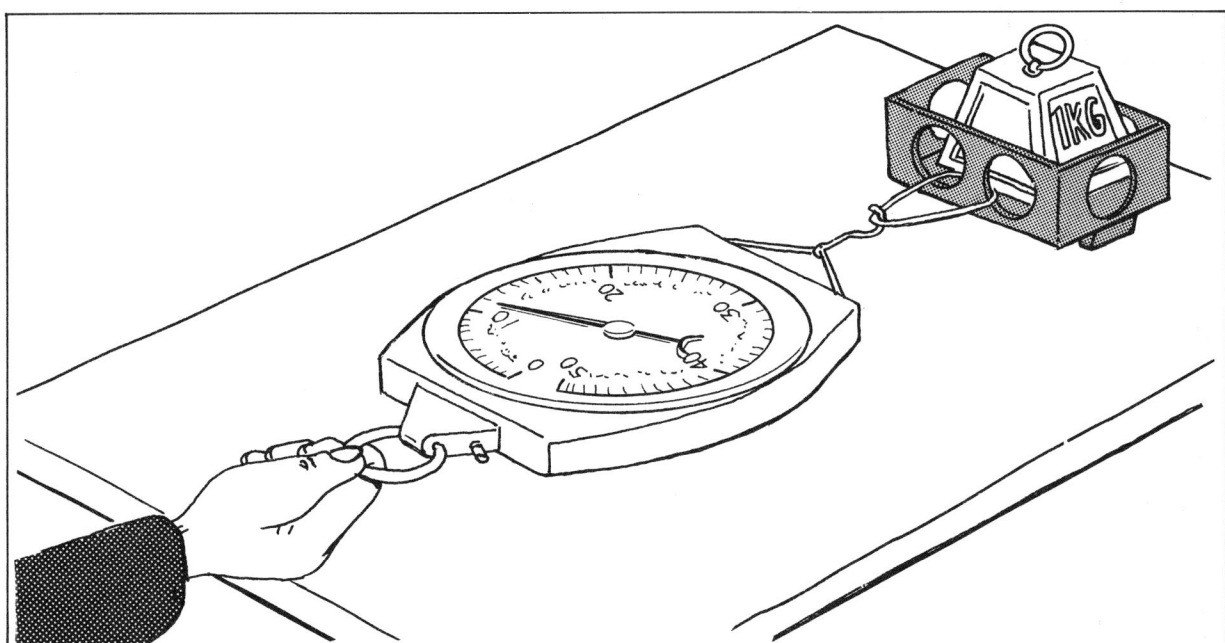

Choose a set of brake blocks and fit them firmly to the brake box.

Place the box on a metal sheet and put a mass in the box.

F S I

Hook a forcemeter to the loop and pull it along the metal sheet. It is not easy to read the forcemeter and you may need to have a few test pulls before you decide on the reading.

Try this test on different surfaces. Use the surfaces dry, then wet.

Try each test twice and get an average result.

Use a table like this for your results.

Braking surface	Force to make blocks slide (newtons)					
	Leather-faced block			Rubber block		
	1	2	Average force	1	2	Average force
Aluminium (dry)						
Aluminium (wet)						
Steel (dry)						
Steel (wet)						

Leave the equipment and bench clean and tidy. **Wash your hands after you have cleaned up.**

Write up your experiment fully.

In your conclusion use your results to say what effect wet weather has
a on all rubber brake blocks and
b on leather-faced brake blocks

Your results might not be what you expected. See if you can think of a better way to test out brake blocks. Write this in your conclusion.

Extension work

Check the bar chart on display. Ask for a piece of graph paper and draw a bar chart using your own results and use this in your writing up.

© N.G. Furnell, 1988. Blackwell Education, Oxford

Summary

A crystal garden is set up and the growth rates of different crystals under different conditions are measured and compared.

Care must be taken with the chemicals used here.

Equipment

- Safety glasses
- Water glass (Philip Harris S80110/5)
- Crystals. Any of the following metal salts are suitable: cobalt (II) chloride (Philip Harris S30800/0), copper sulphate (Philip Harris S32625/7), iron (II) sulphate (Philip Harris S47285/7), manganese (II) chloride (Philip Harris S52895/4)
- Plastic tubes: clear, 50 × 13 mm in diameter. Some chemist shops can supply these quite cheaply. Glass specimen tubes could be used as an alternative (Philip Harris C76360/6)
- Test-tube stand (optional, Philip Harris C78780/5). Some support for the tubes is helpful but Plasticine or small jars are good alternatives
- Tweezers
- Thermometer
- Clock with second hand
- Containers to act as water-baths
- Cleaning brush
- Wiping-up cloth
- Hot water and ice may be required

Water glass is caustic. Safety glasses should be worn *throughout* the experiment. Children should be shown how to wash their eyes with an eye wash bottle or using a tap. Even if diluted, splashes of water glass onto skin or clothes should be washed immediately using cold water. Additional instructions may be on the water glass container.

Crystals may be toxic. Safe practice should always be encouraged. The children should use tweezers for handling the crystals.

Notes

When a crystal of a metal salt is dropped into water glass, branch-like growths appear from the crystal. The silicate reacts with the crystal to form the silicate of the metal as tiny tubes. The colour of the growths is the same as that of the crystal.

The extent of crystal growth is affected by crystal type, crystal size, solution strength, and solution temperature. Children can investigate each of these factors in an open-ended way.

WATER GLASS

Water glass is a solution of sodium silicate. It can be purchased from chemist shops. It is quite expensive and needs to be diluted by adding water (tap water is fine) to give a density of 1.1 g/cm^3. This can be achieved using the formula: **Volume of water to be added = 10 × number of grams of water glass − 11 × number of cm^3 of water glass.**

However, growth still occurs at half and double these concentrations and if a reduced volume of water is added initially (say 50%) it can be further diluted as part of the experiment.

How does your crystal garden grow?

Your job is to find out how fast the crystals grow. Test them in different ways to see if you can find the best method.

You need

- Tubes
- Safety glasses
- Water glass
- Measuring cylinder
- Crystals
- Tweezers
- Thermometer
- Clock with second hand
- Pots
- Cleaning brush
- Wiping-up cloth
- Hot water
- Ice

Water glass is a strong chemical. Wear safety glasses. Wash any splashes off skin or clothes. Wipe up any spills.

F S I

Try this first.

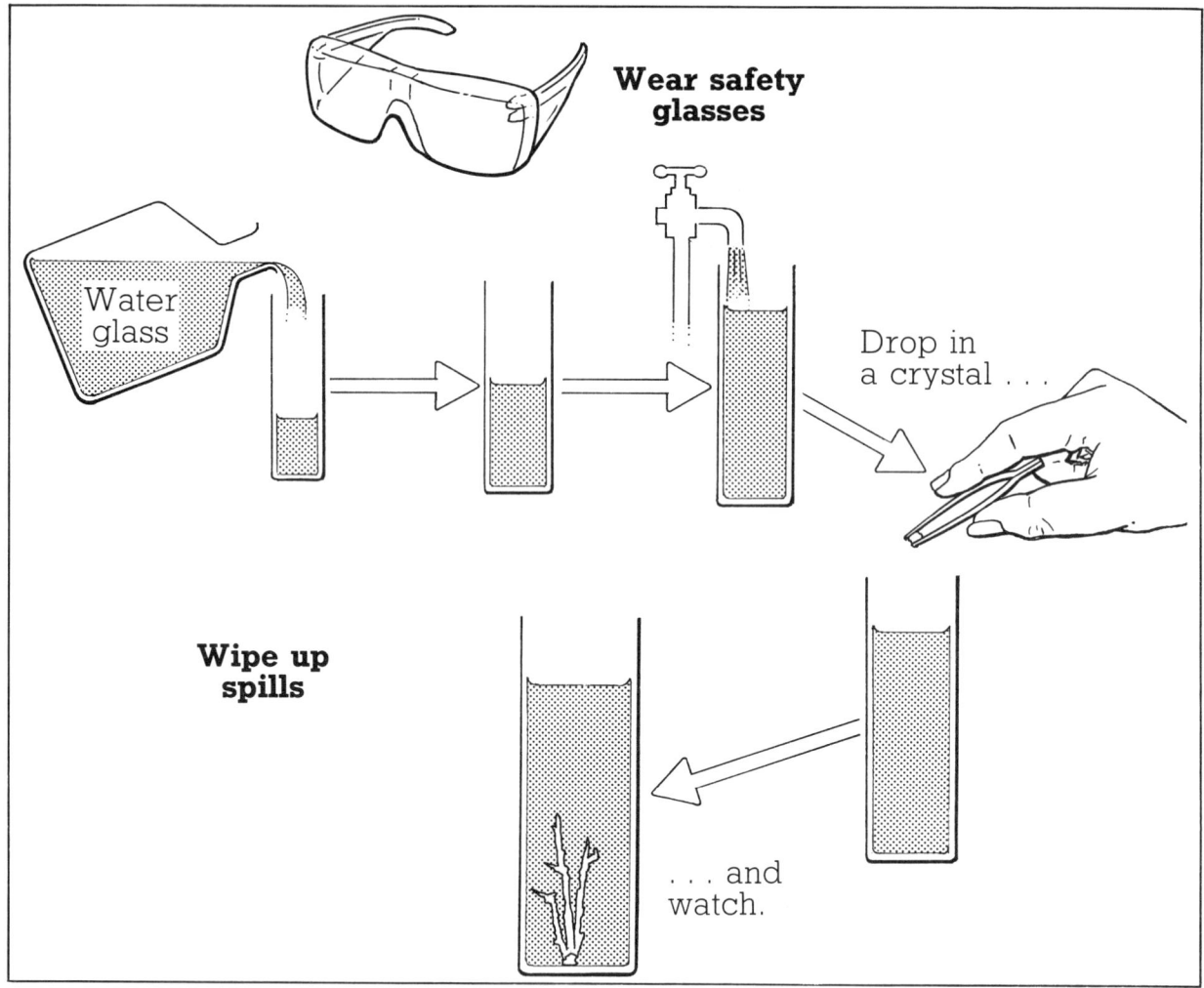

Wear safety glasses

Water glass

Drop in a crystal . . .

Wipe up spills

. . . and watch.

Grow three crystals at a time (in separate tubes) in case some fail!

There are lots of things you could try like:
a using different sorts of crystal and
b using different sizes of crystal.

For each way you try, make a record like this:

Type of crystal _____
What you did for the test _____
_____
_____
_____

                        Tick here
Growth   Good   _____
             Average _____
             Poor   _____
Height grown in 1 minute _____ mm

F S I

Use the cleaning brush to wash out the tubes.
Pour used water glass down the sink.

Write about what you did and what happened
in your experiment. In your conclusion say
what you think makes a crystal grow best.

Extension work

Carry out tests to get the results so that you
can draw graphs with **Crystal growth** (on the
y-axis) and **Time** (on the *x*-axis).

Do this for
a Different crystals on the same graph
b One type of crystal at different
  temperatures
c One type of crystal in different strengths of
  solution

Summary

Adhesive tapes are tested to find the strength of the tape and the strength of the adhesive. The best overall tape is found.

Equipment

- Rolls of masking tape, insulating tape, Sellotape, gummed paper tape, and parcel tape
- Wood block 5 × 4 × 2 cm (as in Nuffield materials kit)
- Retort stand clamp and boss (strong enough to support at least 6 kg)
- Bucket or scale pan (strong enough to support at least 6 kg)
- Assorted gram and kilogram masses (sufficient to break the strongest tape), a total of around 6 kg in 1 kg, 500 g and 100 g sizes
- Scissors that cleanly cut the tapes
- Cloth (or foam block) to cushion the fall of the masses

Notes

This is a two-part investigation using five types of adhesive tape. The investigation can be carried out in various ways. One method is outlined below.

Children should not work without shoes on or use the apparatus near the edge of a bench. Heavy masses falling from bench height will damage floor tiles. Some children prefer to work at floor level.

The five types of tape suggested are commonly available (W.H. Smith sell them all at nearly identical widths of around 20 mm). The packaging and prices need to be displayed for the extension work.

Some tapes can support several kilograms. Do not use tapes that support more than 10 kg otherwise the bucket will probably break!

If the clamp jaws are sharp the tape can be fastened to the 'rod end' of the clamp.

A small (1 gallon) household plastic bucket is ideal.

Gummed paper tape needs a few minutes to dry after it is wet. Having two buckets and wood blocks enables one set to be used for the paper tape and left for the gum to dry while the other tapes are tested.

BREAKING STRAIN

The breaking strain of the tape is found by attaching a plastic bucket to a stand with a *single* length of tape secured by overlapping it slightly (this is illustrated on the pupils' page). Masses are then put in the bucket until the tape snaps. Sometimes children inadvertently overlap the tape so that it is doubled and then find it hard to snap! Strictly, we should use the units of newtons here (10 × the mass in kilograms) since the tape snaps due to the *force* acting on it.

ADHESIVE STRENGTH

If a group is unable to think of a way to measure the adhesive strength it may be helpful if they can refer to this diagram. It shows one way of testing adhesive strength: by taping the bucket handle to the bottom of the wooden block.

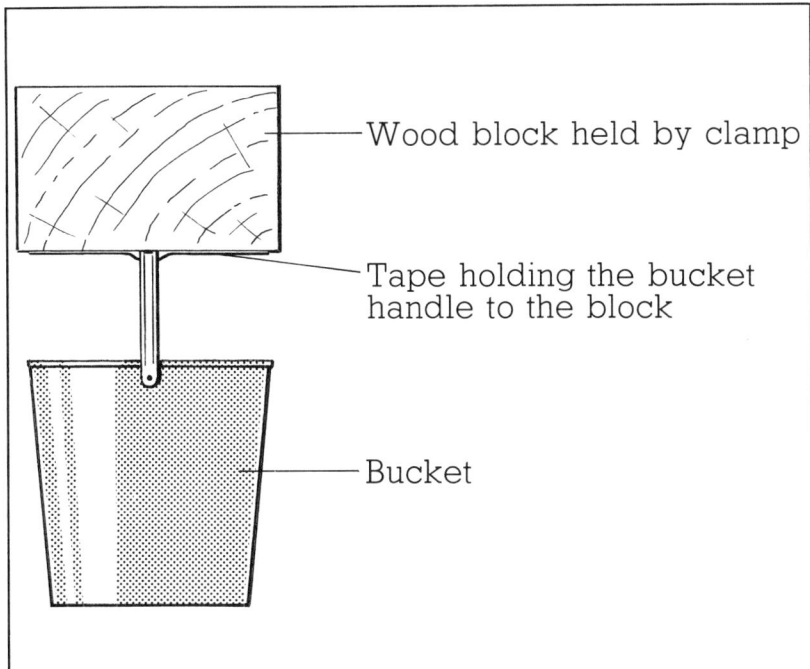

Wood block held by clamp

Tape holding the bucket handle to the block

Bucket

The sticky side of the tape must stay perfectly clean.

Another piece of tape (the same size) can be stuck over the first one. This is to stop the tape tearing and will not alter the adhesive strength. More than two pieces may be needed if the tape tears easily.

Tape aid

Which kind of adhesive tape is best for general use?

Your job is to put some common tapes to the test.

TAPE STRENGTH

How many kilogram masses will a tape hold before it snaps?

See if you can find out for each tape.

Use a table like this for your results.

Type of tape	Mass in bucket (grams)				
	When tape snaps	Score (1–5)	When tape pulls off	Score (1–5)	Average score

When you have tested the five tapes score your results on a 1–5 scale (5 for strongest, 1

© N.G. Furnell, 1988. Blackwell Education, Oxford

for weakest) and put your scores in your table.

ADHESIVE STRENGTH

How sticky are the tapes you have?

See if you can think of a way to test them out. You can use the things you have already and a wooden block.

Put all your results in the table and score them on a 1–5 scale as you did before. See if you can find the average scores and put them in the table.

Write up your experiment using the table to show your results. In your conclusion list:
a The strongest tape
b The tape with the strongest adhesive
c The best tape overall

Also, the adhesive test was for the tapes stuck to wood. Why might this make the test unfair? Explain in your conclusion how you could make the test fairer.

·Extension work

Use the information on the tape packaging to calculate the cost per metre of each tape. Score this as before and find new average scores for each tape.

Does this make any difference to the best tape overall?

Add any new information in your writing up.

If you are able to, try out your ideas for making the tests fairer.

Pressure

T14

Summary

Various everyday pressures are calculated by measuring the forces and areas involved.

Equipment

- Platform scale (Griffin XBY-500-U)
- Spring balance, compression (Philip Harris P11740/2)
- Plasticine (about 100 g)
- Eraser
- Press stud (one on a purse or pencil case is suitable)
- Drawing pin, with a piece of board (10 × 10 × 1 cm approx.) to push the pin into. Use softwood, softboard or plasterboard
- Small mammal: something of a reasonable size as a guinea pig, rabbit or rat is easier to handle
- Calculator
- 1 cm squared paper or graph paper

For extension work
- Chair
- Bed spring
- Cartesian diver
- Hazel nut
- Desk stapler
- Balloon

Notes

Children carry out a variety of activities all of which involve making two measurements to determine pressure. Scales are used as a forcemeter to measure pushing force and the area being pushed on is measured using squared paper.

Pressure is defined as **force/area**. Force is measured in newtons (N) and, for convenience, area is measured in square centimetres (cm^2).

For most of these activities the force involved can be found by pushing down onto (or standing on) the scales. For the Cartesian diver the scales will need to be held sideways to measure the force. Both the 'bathroom' type and 'kitchen' type suggested have newton scales. A kilogram scale is less satisfactory and can be confusing but a newton scale (10 × the kilogram reading) can be written on top of the kilogram one.

Teacher's Notes

The areas can be found by drawing outlines on 1 cm squared paper or graph paper. Very small areas (such as a drawing pin tip) are best estimated by comparison with 1 mm squared graph paper. The pressures can then be calculated in newtons per mm^2 or multiplied by 100 to convert them to newtons per cm^2.

The contact area of both the Plasticine and balloon will change as they are squashed. The applied force is also likely to change. For simplicity the largest forces and areas can be considered.

The contact area of hands and feet is not the same as the area of their outline on the graph paper. A better estimate would come from hand prints or footprints using paints.

A mammal's footprint can be taken as an impression in soft Plasticine or clay.

The press stud and drawing pin both have alternative answers. The area measured can be of the head or of the tip. The head, being larger, is easier to measure.

For the stapler, the area pushed by the hand, or the tips of the staple, can be measured. Some children may ask about this and it may be instructive to explain to them that these differences in pressure are the reason that a drawing pin can be pushed into a board. An explanation of this could involve the following examples.

THE STILETTO HEEL AND THE ELEPHANT

It has been said that an elephant could walk across a linoleum floor with no ill effect but a woman walking across the same floor wearing shoes with stiletto heels could dent it! The reason is that when the woman puts her weight on the heel it is acting on a very small area. This causes a very large pressure on the floor under the heel.

The same principle applies to the drawing pin. Say the head of the drawing pin has an area of 1 cm^2 and the tip area is 1/100 cm^2.

If the pin is pushed with a force of 20 newtons, the pressure is given by: **force/area** $= 20/1 = 20$ N/cm^2, which is the pressure applied to the pin by the person pushing it; but at the tip the pressure is $20/(1/100) = 2000$ N/cm^2, which is the pressure exerted by the pin tip when it is pushed into a board.

THE CARTESIAN DIVER

This is a model that demonstrates the principle of the submarine. A used ink cartridge in a plastic lemonade bottle full of water is ideal.

The cartridge is weighted with a paper clip and partly filled with water so it *only just* floats (open end down) in the bottle. The bottle is completely filled with water and the stopper put on. Squeezing the bottle makes the 'diver' sink. The squeezing increases the pressure, drives water into the cartridge and makes it less buoyant. The more air left in the cartridge initially, the harder the bottle has to be squeezed to make the diver sink.

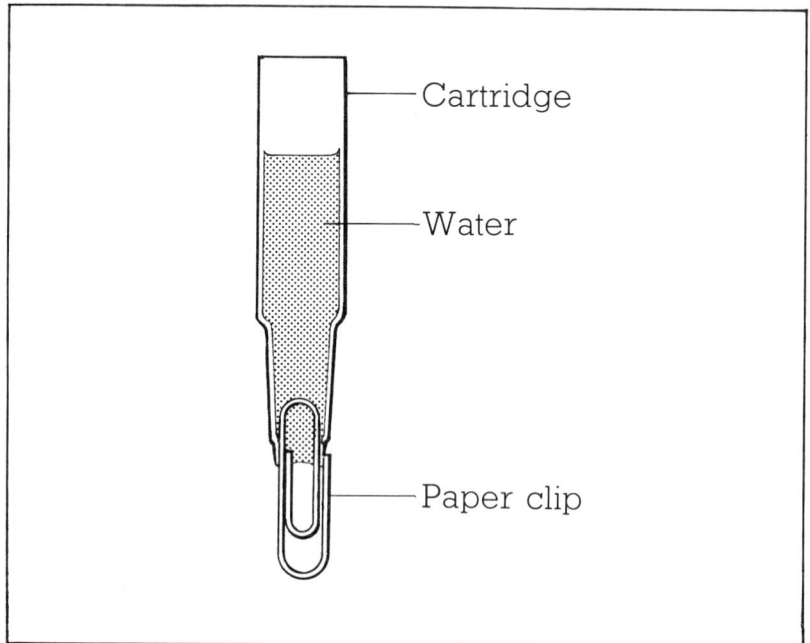

Cartridge

Water

Paper clip

Pressure

When you press on something you are putting pressure on it.

You need

- Forcemeter
- Plasticine
- Rubber (eraser)
- Press stud
- Drawing pin
- Small mammal
- Calculator
- Squared paper or graph paper

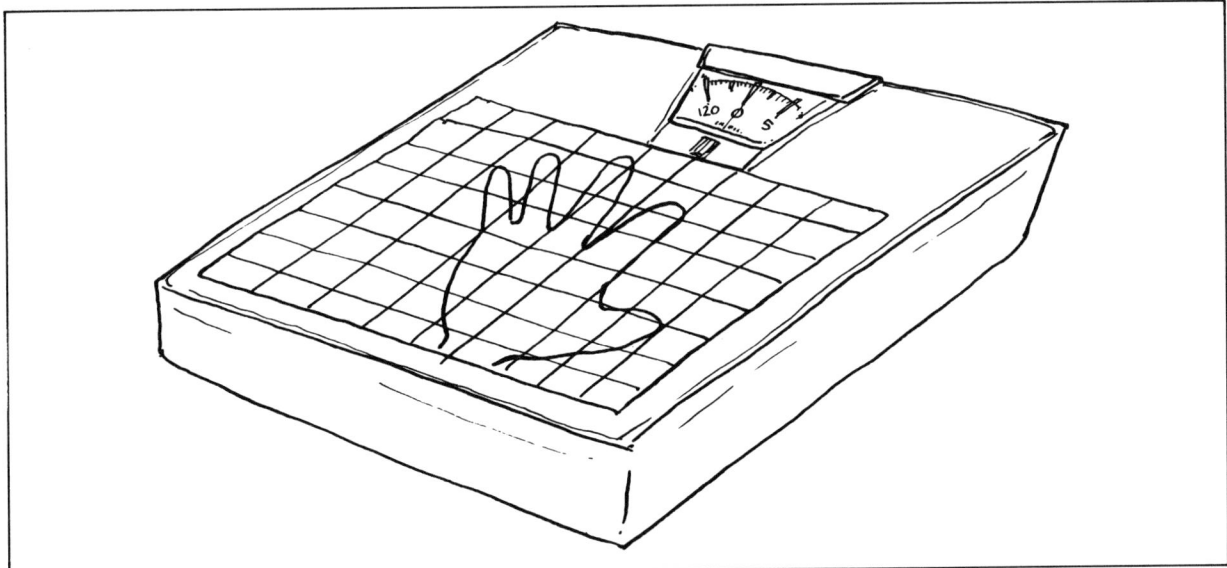

Calculate your hand pressure.

a First measure the **force** you press with (in newtons).

b Then measure the **area** your hand presses on (in cm^2).

c Use a calculator to find the **pressure**. Type in the force (**a**), then press share ($\div$), type in the area (**b**), then press '$=$' to get the pressure (**c**).

F **S** **I**

Put your results in a table like this.

Job done	(a) Force used (newtons)	(b) Area used (cm^2)	(c) Pressure used (N per cm^2)

Using the things you are given see if you can find the pressures needed to do some of these jobs:
- Support yourself standing up
- Squash a ball of Plasticine
- Use a rubber
- Push a press stud shut
- Push in a drawing pin
- Support a small mammal

For each job:
a use squared paper to measure the area being pushed on and
b use the scales as a forcemeter to measure the pushing force.

Write about what you did and use the table for your results. In your conclusion say:
a which job needed the lowest pressure and
b which job needed the highest pressure.

Extension work

Can you think of a way to measure hand or foot area more accurately than drawing around the outline?

© N.G. Furnell, 1988. Blackwell Education, Oxford

F S I

See if you can find the pressures needed to
do these jobs:
- Support a chair you are sitting on
- Squash a bed spring
- Move a Cartesian diver
- Crack a hazel nut
- Use a stapler
- Burst a balloon

In your conclusion say how you would change
this work to make it more accurate.

Summary

Temperature–time graphs are plotted for containers filled with hot water and covered in various materials.

Equipment

- Five empty ring-pull cans (125 ml fruit juice cans are ideal)
- Squares of materials to cover cans such as aluminium foil, white cotton, black cotton, wool, paper, nylon
- Electric kettle (or other means of heating water)
- Pouring funnel that fits into the cans
- Five thermometers that measure from – 10 to 110°C (Philip Harris C79337/4)
- Clock or stopclock
- Elastic bands
- Five stirring rods (pieces of wire coat hanger are fine – use 14 cm lengths with 2 cm bent at 90° each end)
- A5 graph paper
- Tray to stand cans in

Notes

This investigation illustrates the insulating properties of various materials. Five empty drink cans are used, four of them with coverings on the base and sides. Each can is carefully filled with hot (not boiling) water by the teacher. The same volume of water must be put in each can and the coverings must stay dry. It helps to use a pouring funnel. Every 2 minutes the temperature of each can is recorded. The water should be well stirred before taking readings.

Avoid letting children cut aluminium foil from a roll – they are not very tidy about it!

A wall clock with clear minute markings and a second hand is recommended. One end of 150 cm thread can be tied to each thermometer and the other end attached securely to the wall above the workplace. This will reduce the likelihood of breakage.

Putting on the tea cosy keeps the teapot hot but does it matter what material the tea cosy is made from?

Your job is to find the answer to this question.

1 Choose some materials you want to test and fit them round the cans.

2 Use a results table like this. Write in the names of the coverings.

Covering on can	Temperature (°C) taken every 2 minutes										
	0	2	4	6	8	10	12	14	16	18	20
No cover											

3 You need to have your thermometers, stirrers and clock ready. As soon as you have had the water put in the cans you need to start getting results.

4 Ask for the cans to be nearly filled with hot water. **Do not wet the coverings.**

5 Carefully put a thermometer in each can.

6 Read them as soon as they stop rising and put the results in your table.

7 You need to take the temperature of the cans every 2 minutes. Stir the cans before each reading.

8 Get results for 20 minutes.

9 Plot a graph of your results for each can on the same piece of graph paper (see the graph on display). This type of graph is called a **cooling curve**.

10 Write up your experiment fully using your cooling curve graph to present your results. In your conclusion say:

   a which of the coverings tested would be best for a tea cosy,

   b which covering was the best insulator,

   c if you think the colour of a covering makes any difference.

Summary

Brands of toilet soap are compared for value, appeal and cleaning power and the 'best buy' found.

Equipment

- Ten bars of soap (two each of five different brands)
- Sink or bowl
- Water supply
- Towel or paper towels
- Balance
- Pot of shoe polish
- Pot of 'grime'
- Bottle of ink or ball pen
- Marker pen (permanent type)
- Five plastic bags

Notes

This is a test of the cleaning action, appeal and value for money of five brands of soap. Children mark their fingertips with the ink, polish and 'grime' provided and then use the soaps to wash their hands. (This is repeated for each brand.) Unused bars are weighed to find the cost per gram and compared for their appeal. Finally, the 'best buy' is chosen by combining the results for each test.

The sample of children's work which has been evaluated (see pages 11–12) shows clearly that a calculator is required for the calculations involved.

A soap's cleaning ability could be compared with washing in water alone, but the criteria of 'appeal' and 'cost per gram' do not relate to water.

Children may want to time the stages of this experiment. A wall clock with a second hand is adequate.

It is useful to use bars of different coloured soaps. For variety you can include a household soap, a medicated soap, a toilet soap and one specially for sensitive skin. You may like to broaden the range and include a liquid soap and a 'Swarfega'-type hand cleanser.

One bar of each brand of soap should be kept unused in a sealed plastic bag so that the mass, shape and colour of new bars can be referred to.

T16

The wrappers and prices of each brand need to be on display.

The balance really needs to weigh to the nearest gram (or small multiples thereof). This is particularly true for the extension work. If none is available choose brands of soap that have the mass of the bar printed on the wrapper.

Do you use the best soap?

Is the same brand used by film stars, pop stars, you and your best friend?

Your task here is to find the best of the bars!

You have five bars to test out.

CLEANING POWER

1 Your second job is to test each soap on dirty skin . . .
Your first job is to get some mucky fingertips to wash!

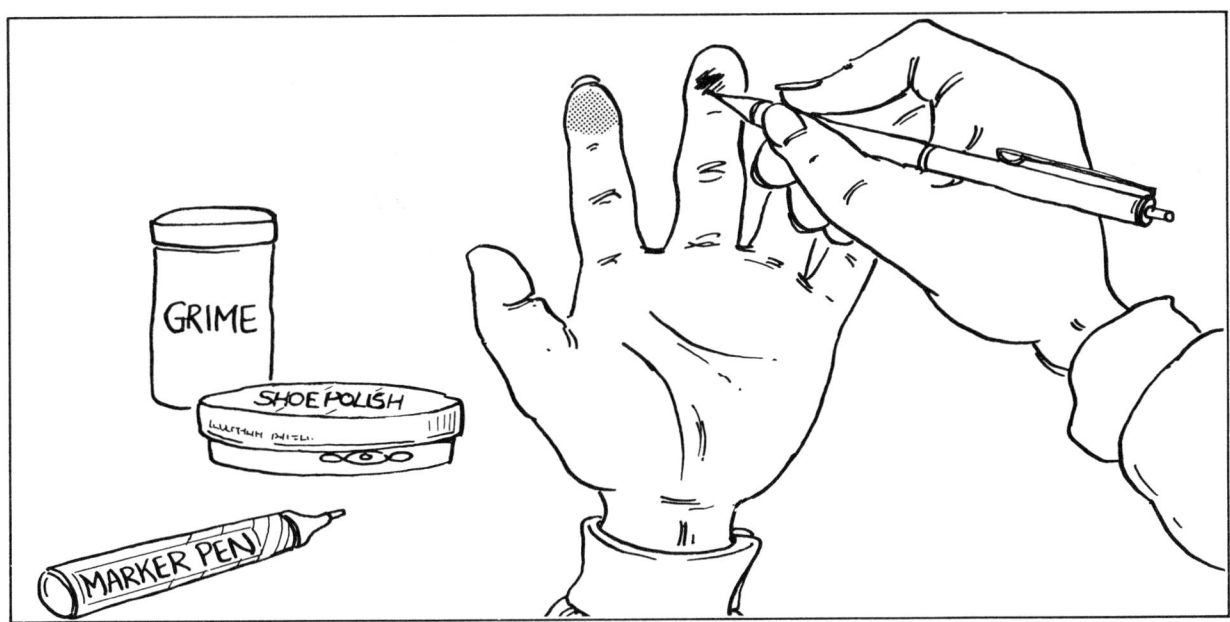

2 Use a table like this for your results.

Brand of soap	Scores for cleaning (1–5)				
	Grime	Polish	Felt pen	Ink	Average

3 Give scores out of 5 for each soap's cleaning power.

F **S** **I**

4 See if you can find the averages using a calculator.

APPEAL

What things make soap nice to use?

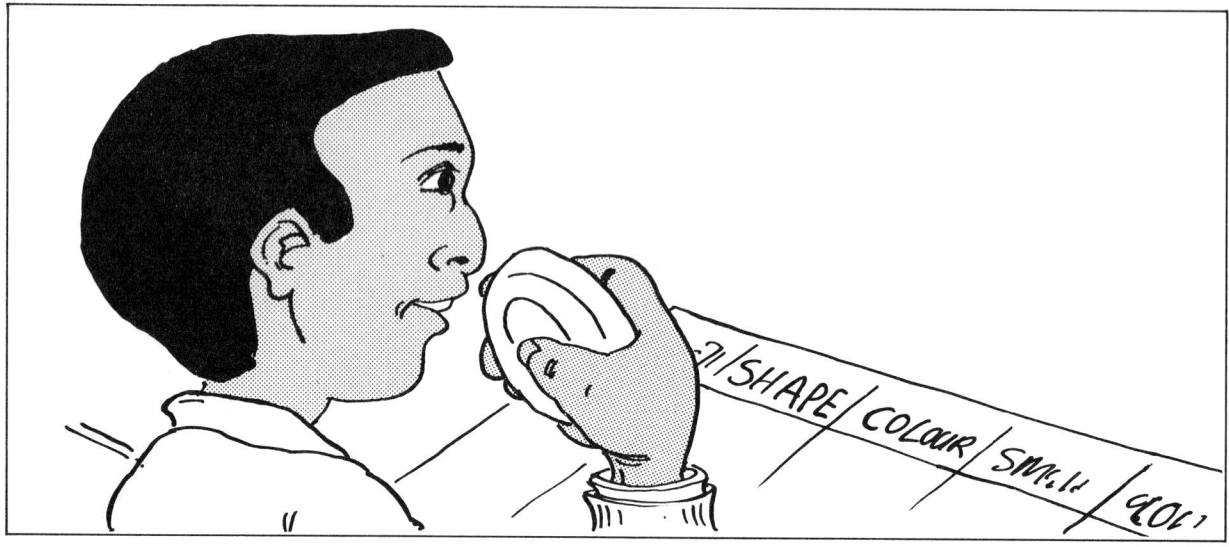

5 See if you can think of four ways to test the soap's appeal.

6 You can use the same type of table again – just alter the headings.

COST

Your last job is to check out the cost per gram of the soaps.

7 Use a table like this for your results.

Brand of soap	Soap costs			
	Price (p)	Mass of bar (g)	Cost per gram (p/g)	Score (1–5)

8 Find the price from the soap wrappers – use **new** bars to find the mass.

F S I

9 Use a calculator to find the cost per gram. Type in the price, then press share ($\div$), type in the mass, then press '='.

10 Score the soaps on a 1–5 scale (5 for the cheapest per gram, 1 for most expensive).

11 See if you can find the average scores.

FINAL SCORES!

Now you have matched the soaps, you need to get the aggregates (totals) of the scores!

12 Use a table like this and carefully fill in all the scores.

Brand of soap	Scores (1–5)			Aggregate score	Average score
	Price	Cleaning	Appeal		

13 Find the overall average scores using a calculator and write these in the last section.

14 Write up your experiment using the first and last tables for your results. In your conclusion list the soaps in order of their final scores. Write about any results that surprised you!

Extension work

An important point about soap is the amount used each time.

1 See if you can think of a fair test to estimate how many 'hand washes' each bar of soap would give in total. **Your method should not use more soap than the cleaning test.**

2 If you are able to, try your ideas out.

© N.G. Furnell, 1988. Blackwell Education, Oxford

Summary

A hand cream is made up and the mixture progressively altered in order to improve on the original.

This activity is suitable for children who can safely use hot water.

Equipment

- Lanolin (Philip Harris S45860/8)
- White petroleum jelly (Vaseline)
- Perfume
- Three spatulas or small scoops (Philip Harris C68345/3)
- Plastic glue spreader for use as stirrer/scraper
- Six 10 ml beakers (Philip Harris C16621/5)
- Washing-up bowl (for water-bath)
- Paper towels or wiping-up cloth
- Large tray
- Hot water
- Marker pen

Notes

A basic hand cream can be made from an emulsion of lanolin, petroleum jelly and water. A suitable emulsion has a lanolin : petroleum jelly : water ratio of approximately 2 : 3 : 2 (by volume).

As a starting point for the activity, three level scoops of each substance could be used. (This is likely to produce a poor emulsion.) Further recipes might involve (1) using less lanolin (likely to produce a poor emulsion), (2) using less petroleum jelly (no emulsion likely), or (3) using less water (emulsion achieved but may be rather greasy). The final batch is made according to previous results and a little perfume can be added.

Using a washing-up bowl as a water-bath gives plenty of room to work in. The water needs to be quite hot (50°C) to melt the mixture. Hot water from the tap is often hotter than this and quite adequate.

Do not allow groups to heat mixtures directly over a flame or hot surface.

Children must take care to clean their hands after handling the apparatus and materials. It may help to use scrap paper for written work during the experiment – books can get very greasy!

A large tray and a clean rinsed cloth (or paper towels) for wiping up are important. Unless carefully supervised, children can quickly make the workplace messy and the floor slippery.

So long as the amounts made up are small, the activity is reasonably economical.

TESTING

Factors that can be considered here include:
- Cost per gram (groups may ask how much the constituents cost by weight)
- Ease of application
- Greasiness after application
- Resistance to washing off
- The cream remaining an emulsion (In some cases the water in the hand cream separates out after a short time.)

Most Vaseline and lanolin combinations will soften the skin and comparing recipes may not reveal much difference between them.

Colour and scent need not be considered here, but they can be added as desired.

Hand cream

What makes a good hand cream?

Your job is to see if you can make the 'ideal' hand cream – a pleasant creamy mixture.

You will need to mix lanolin and Vaseline (petroleum jelly). It is best to put them into a beaker that is standing in warm water so they melt.

If you are going to add water do this last, using hot water. **Stir as you add it** so it mixes with the lanolin and Vaseline.

Measure things out in level scoops and write a 'recipe' for each one as you make it.

When your first mixture is made, try it out and give it a score for all the things you think are important about a hand cream.

Test small amounts on the back of your hand. Is it too greasy? How much does it cost per gram? See what ideas you can think of for testing your hand cream.

Make new 'recipes' and try them out as you make them. Make up six different types.

 © N.G. Furnell, 1988. Blackwell Education, Oxford

F **S** **I**

When you have finished, scrape any unused hand cream onto a paper towel and fold it up before throwing it away. Use paper towels to wipe the beakers and spatulas clean. Leave the work place clean and tidy. **Now give your hands a good wash!**

Write up your experiment fully. Give the recipes and their scores in your results. In your conclusion put all the things you think are important about a hand cream. Say what things made your best hand cream your first choice.

Summary

Models of various shapes are made up and tested in a tank containing male guppies.

Equipment

- Tank of 5–10 male guppies
- 100 g of Plasticine
- Thread or wire
- Scissors or wire cutters

Notes

Classic research into stickleback behaviour was carried out by Niko Tinbergen using models. (See, for example, A. Manning, *An Introduction to Animal Behaviour*, Edward Arnold, 1967.)

A variety of fish 'shapes' are shown in the illustration on the pupils' page. Children can make these shapes out of Plasticine and support them on a wire or thread. (A model is made more secure by bending the wire at the bottom and moulding the model around it.) The models can be tested by moving them around in the tank to simulate a fish swimming.

It is not intended that guppies be kept permanently for these activities. They are widely available and could be loaned for a short period to carry out the investigation.

SCORING

A table of results is not expected for this work. Here are suggested marks for possible responses.

Result of test	Suggested mark
Model is followed by fish	4
Fish do not react	2
Fish try to eat model	0
Fish are afraid of model	0

Guppies are a convenient choice because they live and breed quite well when kept at room temperature. The females are live bearing and produce several broods a year. Male guppies are often seen pursuing females around the tank and they will follow a model fish if it resembles a female before the eggs have been fertilised. It is helpful if the children can see female guppies when they are making their models.

They have quite a distinctive shape (shown centre left in the illustration on the pupils' page). If one or two of the better models are kept, once they are made, they can be given to groups who find making them difficult. Guppies can be transferred from their aquarium to the test tank as required. This keeps the aquarium clean and relatively undisturbed. They are likely to try and eat anything small – their own fry included!

Other fish that shoal well, such as Tiger Barbs, might be a better choice experimentally but require a heated aquarium.

Occasionally children want to add refinements to their models or use other materials like carrot or aluminium foil. These ideas are acceptable so long as they remain fair tests i.e. if *all* the differently shaped models are made from foil or carrot.

During trials some children found it difficult to make their models move realistically without alarming the fish. Those who used a wire had better control of the model even though it was more obvious. If the light reaching the tank is from the opposite side to the experimenters, fish are less disturbed by shadows and movements above them.

Reference

Animals in schools, November 1986, RSPCA

Guppy models

How good are fish at telling model fish from real fish?

Your job is to make up five models and test them out.

You need

- A tank of male guppies
- Plasticine
- Thread or wire
- Scissors or wire cutters

Use Plasticine and cotton or wire to make up your models. Try different shapes and sizes.

Let the fish test your work!

Test each model in turn. Use the thread or wire to move your model in the tank. Try to make it move slowly and act like a fish.

The best test is to see if the fish will follow your model around.

F S I

Give your models marks for their tests.

Give **low** marks for:
a models that fish swim away from and
b models that fish try to eat.

Give **middle** marks for models that do not bother the fish.

Give **high** marks for models the fish follow around.

Write up your experiment fully. To present your results draw each model and write its score next to it. In your conclusion say:
a if the size of the model makes any difference,
b if female guppy models were more popular than other models, and
c if the fish seemed frightened by your movements.

How could you change the experiment so that the fish would not be disturbed so much? Write about your ideas in your conclusion.

Extension work

Which of these variables is the most important factor in how a real fish reacts to a model?
• Model size
• Model shape
• Model colour(s)

See if you can carry out some tests to find out.

Summary

Orange squash drinks are made up from different brands and the cost per drink is calculated. A taste test is carried out and the best brand selected. A vitamin C test and additives check are included as extension work.

Equipment

Use this equipment for drinks only.

Different children should not drink from the same beaker.

- Six small disposable drinking beakers
- 25 cm^3 measuring cylinder (Philip Harris C31780/1)
- 50 cm^3 measuring cylinder (Philip Harris C31800/3)
- Six brands of orange squash
- Marker pen or sticky labels for beakers

For extension work
- Small container or 100 ml beaker (Philip Harris C17500/9)
- 3 cm^3 graduated dropping pipette (Philip Harris C61502/2)
- 0.1% DCPIP (2,6-Dichlorophenolindophenol, Philip Harris S35210/4)
- Wiping-up cloth

Notes

Children are often experienced makers of orange squash and know when it tastes 'just right'! In this activity they use their talent to test out six different brands of squash.

They are able to find the cost per drink by making up small samples of each squash and measuring the amounts added. The drinks are then used for a taste test and the 'best buy' is chosen according to taste and price.

During trials some children did not measure both the squash and the water when making drinks. If groups produce a work plan which you check before they begin, this can be explained if necessary. (See 'How much to add?', below.)

Food additives like the orange colours tartrazine (E102) and annatto (E160b) may affect some people causing intolerance reactions (*Which?* Magazine May 1986).

To give a varied range of brands of squash, you might include three branded varieties and any supermarket own brands locally available. (Slimmers' squash is likely to contain tartrazine.) Very recently many companies have stopped adding tartrazine or similar artificial colours. Try to find at least one brand containing one or more of the following, and other brands with none of them in.

- Tartrazine (E102)
- Annatto (E160b)
- Quinoline yellow (E104)
- Sunset yellow (E110)
- Yellow 2G (E107)

The values obtained for the cost per drink will be greatly increased if children are extravagant when mixing the drinks.

DISPLAY FOR EXTENSION WORK

Additive information: p. 222, *Which?* Magazine May 1986, Consumers' Association, 14 Buckingham Street, London WC2N 6DS. Teachers may make photocopies for teaching purposes only.

To make up 0.1% DCPIP solution dissolve 0.25 g of the solid in 250 cm^3 distilled water.

HOW MUCH TO ADD?

Determining this can be achieved in one of two ways.

1 The same amount of squash and water can be used for each brand; for example, 20 cm^3 squash and 60 cm^3 water.

2 Squash can be added to (say, 50 cm^3) water until each brand tastes the same strength. Disposable plastic 'cups' can be filled with water up to a 'ring' and squash added in 5 cm^3 lots as required.

The second method involves both taste and dilution (two independent variables) which is a complication. The advantage of it is that the cost per drink figure is based on price and on how the squash might be used. For the first method the cost per drink is dictated by cost alone.

T19

The following information may be helpful for children to have:

If a food company tries out a new recipe they test out the meal on a group of people to see if they like it.

NEW RECIPE CARROTBURGERS WITH GARLIC.

We call this group a **taste panel**.

Taste is not something that can be measured on scales or with a ruler or stopwatch. It gives a qualitative result, and not a quantitative one. **Quantitative results**, where we can measure things like speed or temperature, are more precise because we get values for them. Testing things like wine or perfume or make-up all give **qualitative results** like the taste test. We can make these results more reliable by repeating the test lots of times. That is why we carry out a survey where we ask lots of people the same thing.

Avoid giving children full bottles of liquids which may get dropped or spilled. Ideally keep duplicates and ¼-fill the experimental bottles from the stock bottles as required.

Note that there are distinctions between the terms 'orange drink' and 'orange squash' – see *Which?* Magazine July 1987.

Lovely orange squash – just the job on a hot day!

Your job is to pick the best of the squashes.

HOW MUCH TO ADD?

Use clean containers and mains tap water.

'Dilute to taste', it says on the label.

Make up one drink for each squash and **measure** how much you use. You may need a taste check here – but only a small sip for now!

Pour unused squash away – not back into the bottles.

Have a table like this for your results.

Name of squash	Squash costs				
	Price (p)	Volume (cm³)	Cost per cm³ (p/cm³)	Squash added (cm³)	Cost per drink (p)

Put your results in the table under 'Squash added'.

Check the bottle labels for the price and volume and put them in your table.

With a calculator find the cost per cm^3: type in the price, then press share ($\div$), type in the volume, then press '$=$'.

You can now find the cost per drink. Type in the cost per cm^3, then press times ($\times$), type in the squash added, then press '$=$'.

If you are going to do the vitamin C test keep 10 ml of each diluted squash for later.

TASTE TEST

Now for the important bit!

Carry out a taste test. Plan how to do it first. Remember that drinking one squash straight after another may affect its taste.

Score your results (6 for best taste, 1 for worst) in a table like this.

Squash	Taste (1–6)	Cost per drink (1–6)	Total score

F S I

Now score the cost per drink (6 for lowest, 1 for highest).

Add the two scores to get the total.

Write up your experiment fully – the second table is important for your results. In your conclusion,

a list the orange squashes in order of your results, and

b say why you think it is important in this experiment to dilute the drinks to the right strength.

Extension work

You need tables like these for the next two tests:

| Squash | Vitamin C test | | Squash | Additive check | |
	Amount of squash added (cm³)	Score (1–6)		Additives listed	Score (1–6)

VITAMIN C CONTENT

We can test for vitamin C using a chemical we call DCPIP. It is a blue liquid that turns colourless if you add vitamin C to it.

1 Choose a squash and pour 20 cm³ (undiluted) into a 25 cm³ measuring cylinder.

2 Use a pipette to measure out **exactly** 2 cm³ of DCPIP and put this into a clean 100 cm³ beaker.

F S I

3 Now pour the squash, a little at a time, into the beaker with the DCPIP in. If the squash is rich in vitamin C you will not use more than 10 cm³.

4 Swirl the liquid round to mix it. Stop adding squash when the blue colour disappears.

5 If the mixture still looks dark after all the squash is added it means there is not much vitamin C present – do not add more squash.

6 Check your measuring cylinder to see how much squash you added and write the result in your table.

7 Do this for each type of squash.

8 Score your results – 6 for best (least squash used) to 1 for worst (most squash used).

9 You could also try juice from an orange for comparison.

ADDITIVES

Food additives like the orange colours tartrazine (E102) and annatto (E160b) may affect some people causing **intolerance reactions** (*Which?* Magazine May 1986). If a squash does not contain these we might rate it higher than one that does. If you want to check the full list of suspect additives, it is on display.

10 For each squash, check the ingredients list to see if it contains tartrazine or annatto.

11 If it contains both, give it a score of, say, 0 in your table. If it has one, give it, say, 3. Give it 6 if it has neither.

12 Use a table like this for all the scores. Add these to get the total.

Name of squash	Squash scores				
	Taste (1–6)	Cost per drink (1–6)	Vitamin C (1–6)	Additives (1–6)	Total Score

13 a In your conclusion, write about any difference in the total score you now have and the total score for taste and cost. You may need to use the last table in your writing up.

b Taste, cost, vitamin C content and additives were marked as if they were equally important (out of 6). Try changing the marking so that the things you feel are more important get more marks. Say how this affects the results you get.

c See if you can think of some ways to make this experiment more accurate e.g. only one bottle of each type of squash was used.

F S I

Summary

A flageolet is made from a piece of plastic pipe. The block is adjusted to give optimum sound. A Swannee whistle, nightingale and penny whistle are made up.

This work is suitable for children who can use a junior hacksaw, a Stanley knife and a hand drill.

Equipment

- Pipe: 20 cm or longer. 15 mm 'Pipex' plastic water-pipe (1 928 85) is suitable
- Wooden dowel: 20 cm length with a diameter of just less than pipe bore (approximately 12 mm diameter for the pipe above)
- Thin card
- Plasticine
- Insulation tape
- Glue (e.g. Marvin)
- Beaker of antiseptic
- Beaker for water
- Engineer's vice
- Junior hacksaw
- Craft knife
- Hand drill with 1/8" or 3 mm bit
- Fine file
- Scissors
- Ruler

Notes

A penny whistle is a simple form of recorder and one can be made quite easily from a piece of plastic pipe. Before making holes along the pipe it can be used as a nightingale (warbler) and as a Swannee whistle.

Children should have experienced using the tools required beforehand. Those who are not experienced in sawing tend to work hurriedly without too much regard for precision. The pipe is easy to saw and drill but it must be held securely in a vice.

To cut the mouthpiece it is best to clamp the pipe at the required angle and saw vertically downwards. Children do not automatically realise this and try to use the saw at an angle with varying degrees of success.

It is recommended that the work be tried out beforehand. Explaining how to fit the 'block' in the best position is easy once you have done it yourself!

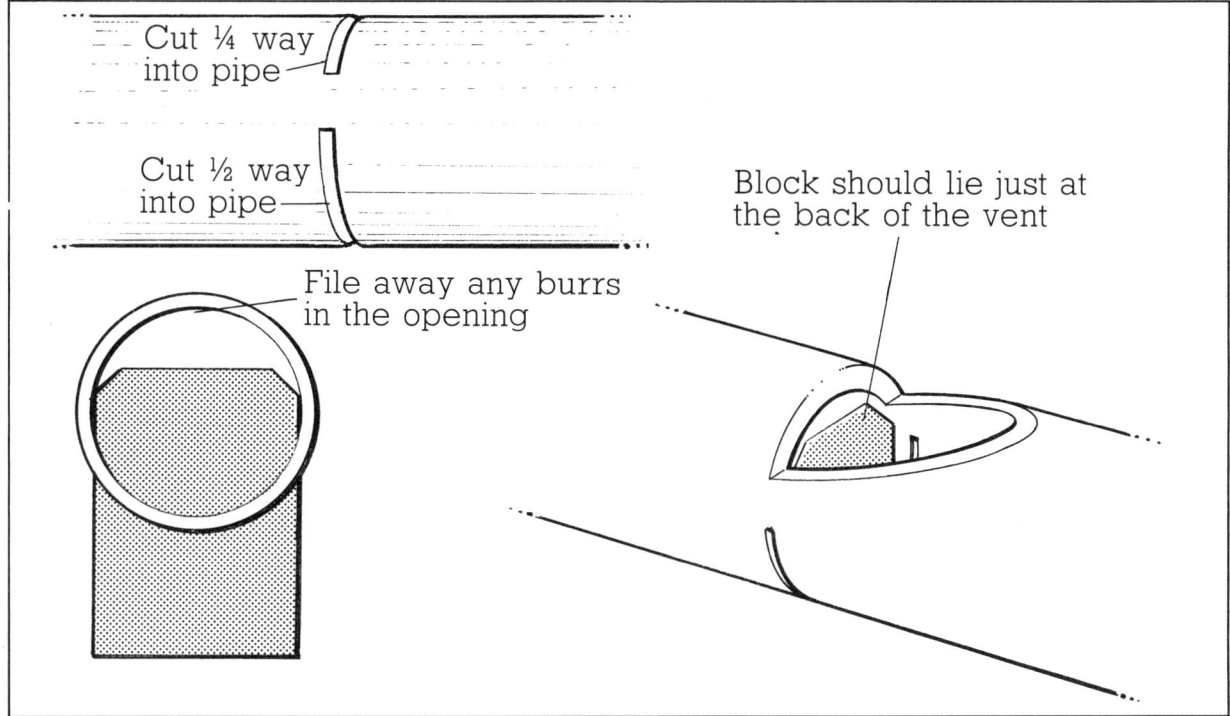

Cut ¼ way into pipe

Cut ½ way into pipe

File away any burrs in the opening

Block should lie just at the back of the vent

Positioning the block

The two saw cuts (step 2) need to be directly opposite so that the block fits just at the back of the vent.

The vent is cut about a quarter of the way into the pipe. If this cut is made too deep it will meet the bottom cut!

The block is pushed into the pipe so that it blocks about three-quarters of the bore. This is best adjusted when blowing down the pipe (with the far end plugged).

The antiseptic can be any suitable preparation such as Milton. For good hygiene practice discourage children from sharing whistles.

Swannee whistle

A piece of wooden dowel makes a suitable plunger. This is moved along inside the pipe as the whistle is blown.

Nightingale

This gives a warbling sound as air from the whistle bubbles through water. The end must be just under the water and it may be necessary to blow harder than normal.

Penny whistle

Suitable positions for holes can be found along the pipe on a trial and error basis. The end of the pipe needs to be partly open to give the bottom note.

Penny whistle

What makes a whistle whistle?

Your job is to make a whistle so you can see
how it works.

**You should only do this work if you can
safely use the tools shown.**

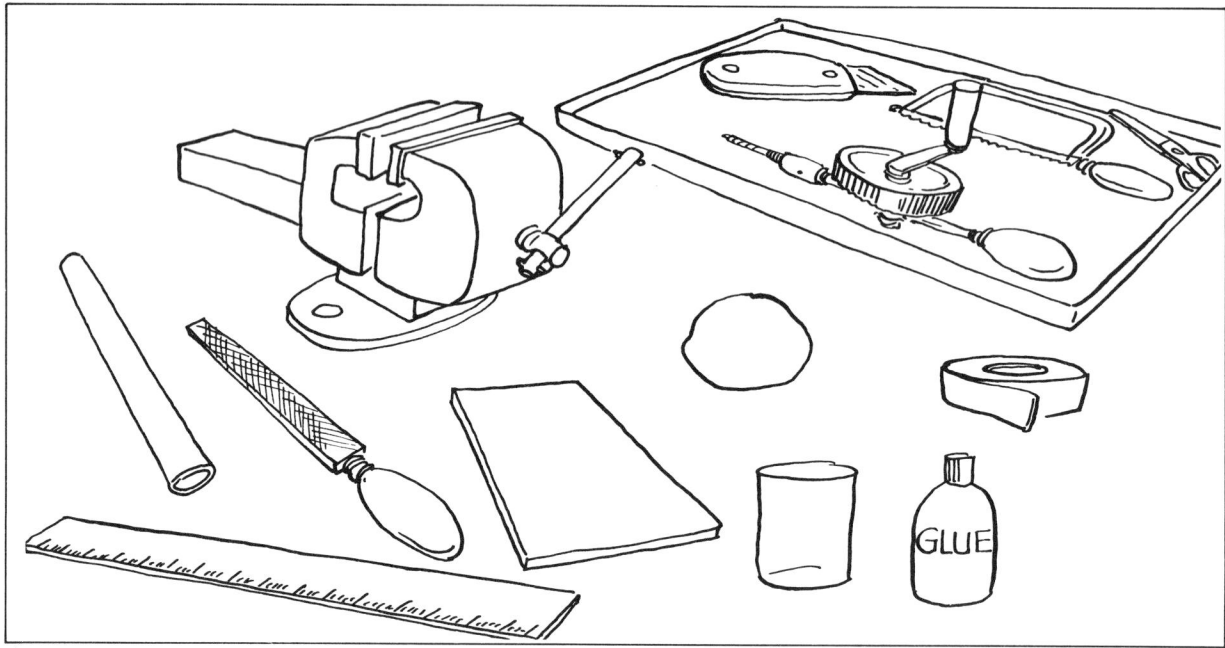

1 Make the body from a piece of plastic pipe.

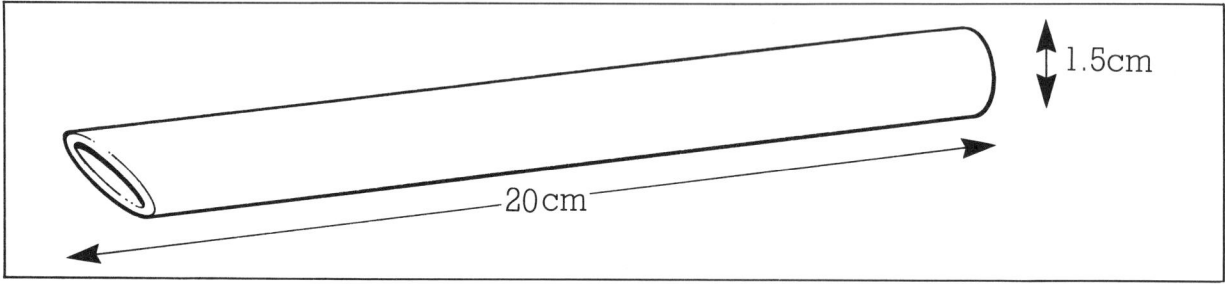

1.5cm

20cm

2 Make neat saw cuts in the top and bottom of the pipe 4 cm from the
slanted end as shown.

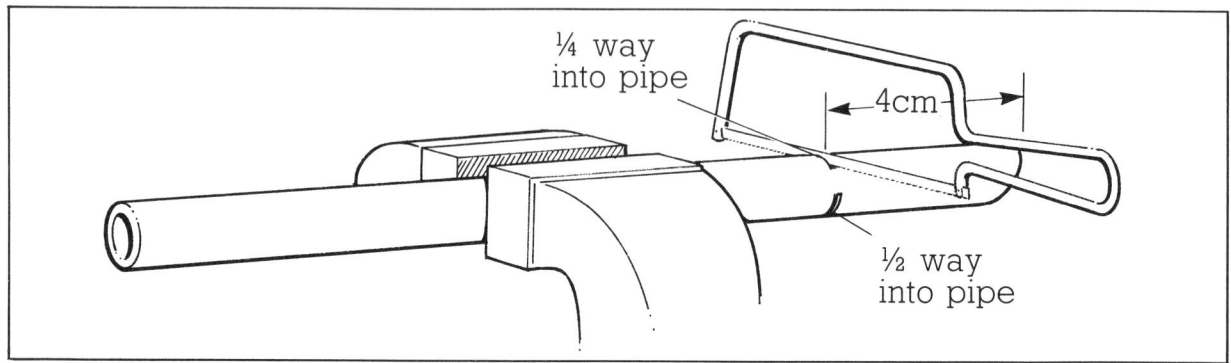

¼ way
into pipe

4cm

½ way
into pipe

F S I

© N.G. Furnell, 1988. Blackwell Education, Oxford

3 Use a craft knife to cut the pipe away . . .

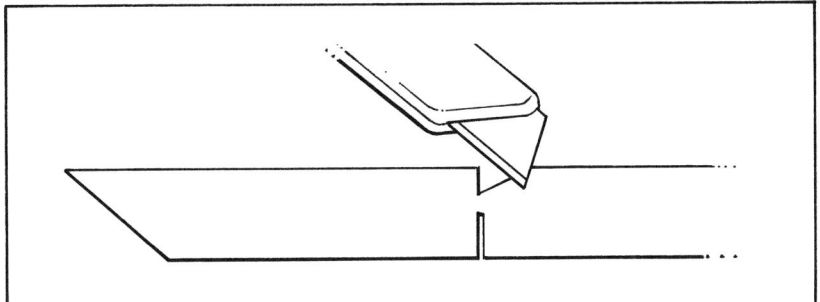

. . . so that there is a small hole.

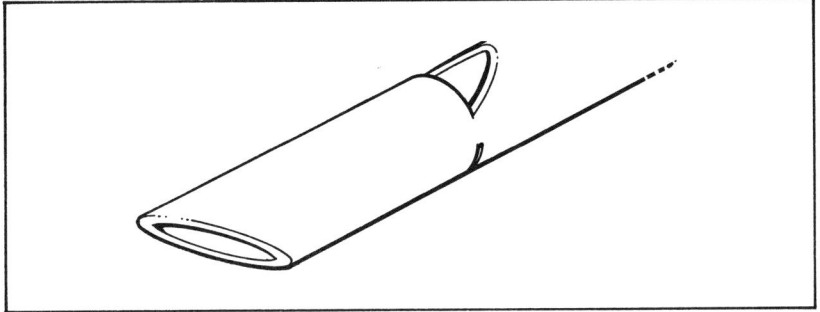

Check against a recorder if you need to.

4 You need a **block** made from thin card.
Trim to just fit into tube.

5 Fit it by pushing it into the slot at the bottom.

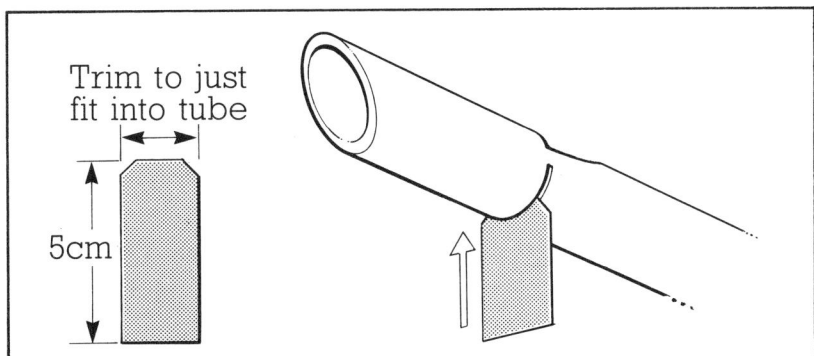

Trim to just fit into tube

5cm

6 Use a file to smooth rough edges. Clean the
mouth piece by dipping it into antiseptic
(then rinsing).

ANTISEPTIC

7 Plug the end with Plasticine and test your whistle by blowing gently.

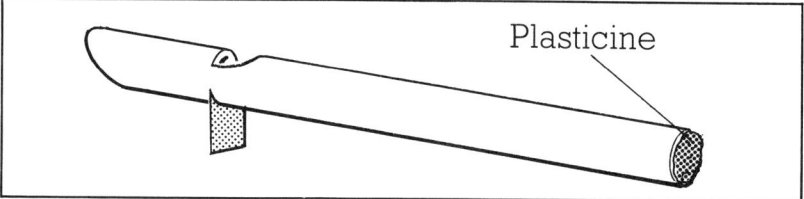

Plasticine

8 Move the block to get the best sound. This can be trimmed and glued in place later on.

Investigations

1 Try out any of these ideas.

a Make a **Swannee whistle**.
Use a sliding **stop** to fit into the body of the whistle.

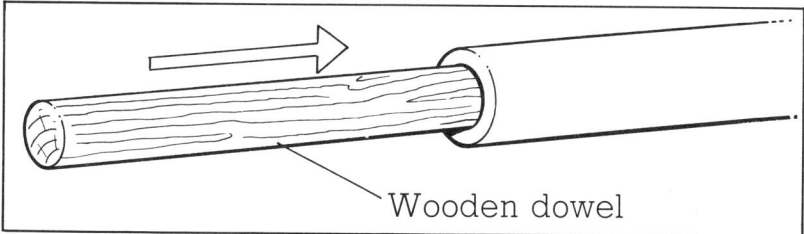

Wooden dowel

b Make a **nightingale**.
Blow with the end of the whistle in water.

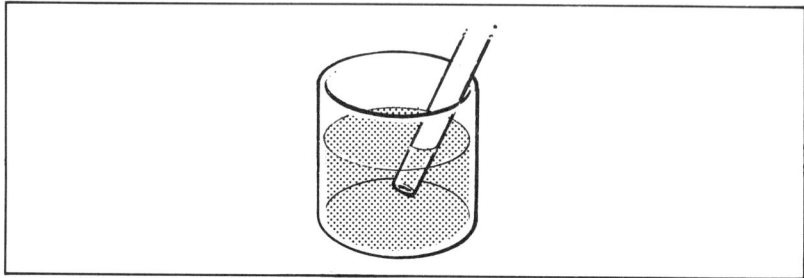

c Make a **penny whistle**.
Drill holes along the body to get a scale of notes. (Tape over any holes you drill and then don't need.)

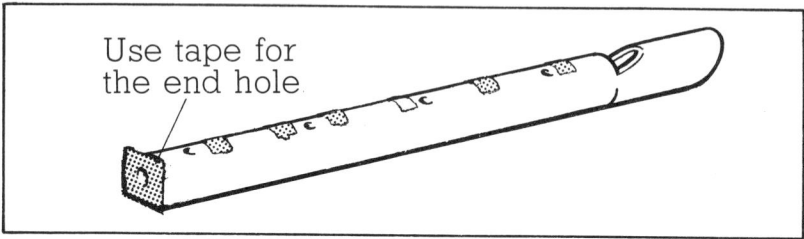

Use tape for the end hole

2 Write about your different investigations.

F S I

Master table sheets

The tables required for investigations are included here. The first two tables have been widely used in the investigations. If it is preferred, children can use these and write in the appropriate headings.

For extension work activities children are expected to draw out the tables for themselves.

Insects landing

Colour of card	Number of insects			
	1	2	3	Average
Red				
Orange				
Yellow				
Green				
Mauve				

F S I

Touch tests

		Objects			

Making-up faces

Make-up styles					
Face A	Face B	Face C	Face D	Face E	No make-up

H's and B's

Pencil symbol	Score 1–5
+	
×	
−	
O	
I	

F S I

Common scents

Scents					
Perfume	**Aftershave**	**Peppermint**	**Lemon**	**Almond**	**Unscented**

Righting wrongs

Correction method	Score out of 10				Average score
	Pencil	**Ball pen**	**Felt tip**	**Ink**	
Ink rubber					
Typists' rubber					
Natural rubber					
Tipp-Ex paper					
Tipp-ex pen					
Eraser pen					

F **S** **I**

A sweet choice

Colour of sweet chosen				
Red	**Orange**	**Yellow**	**Green**	**Purple**

Position on tray of sweet chosen				
1	**2**	**3**	**4**	**5**

Bird feed

Colour	Number of bread cubes		
	at start	at end	eaten
Red			
Orange			
Yellow			
Green			
Blue			
Plain			

F S I

Pet food choice

Type of food	Mass of food (grams)		Amount eaten (grams)
	before feeding	after feeding	

Type of food	Amounts eaten (grams)			
	Pet 1	Pet 2	Pet 3	Average

Best bubbles

Detergent solution strength (%)	Bubble diameter (cm)				
	1	2	3	4	Average
5					

F S I

Bike stoppers!

Braking surface	Force to make blocks slide (newtons)					
	Leather-faced block			Rubber block		
	1	2	Average force	1	2	Average force
Aluminium (dry)						
Aluminium (wet)						
Steel (dry)						
Steel (wet)						

Tape aid

Type of tape	Mass in bucket (grams)				
	When tape snaps	Score (1–5)	When tape pulls off	Score (1–5)	Average score

© N.G. Furnell, 1988. Blackwell Education, Oxford

F S I

Pressure

Job done	(a) Force used (newtons)	(b) Area used (cm²)	(c) Pressure used (N per cm²)

Cooling down – covering up

Covering on can	Temperature (°C) taken every 2 minutes										
	0	2	4	6	8	10	12	14	16	18	20
No cover											

Which soap?

Brand of soap	Scores for cleaning (1–5)				
	Grime	Polish	Felt pen	Ink	Average

Brand of soap	Soap costs			
	Price (p)	Mass of bar (g)	Cost per gram (p/g)	Score (1–5)

Brand of soap	Scores (1–5)			Aggregate score	Average score
	Price	Cleaning	Appeal		

F S I

Orange squash

Name of squash	Squash costs				
	Price (p)	Volume (cm³)	Cost per cm³ (p/cm³)	Squash added (cm³)	Cost per drink (p)

Squash	Taste (1–6)	Cost per drink (1–6)	Total score

F S I

Bar charts

Touch tests

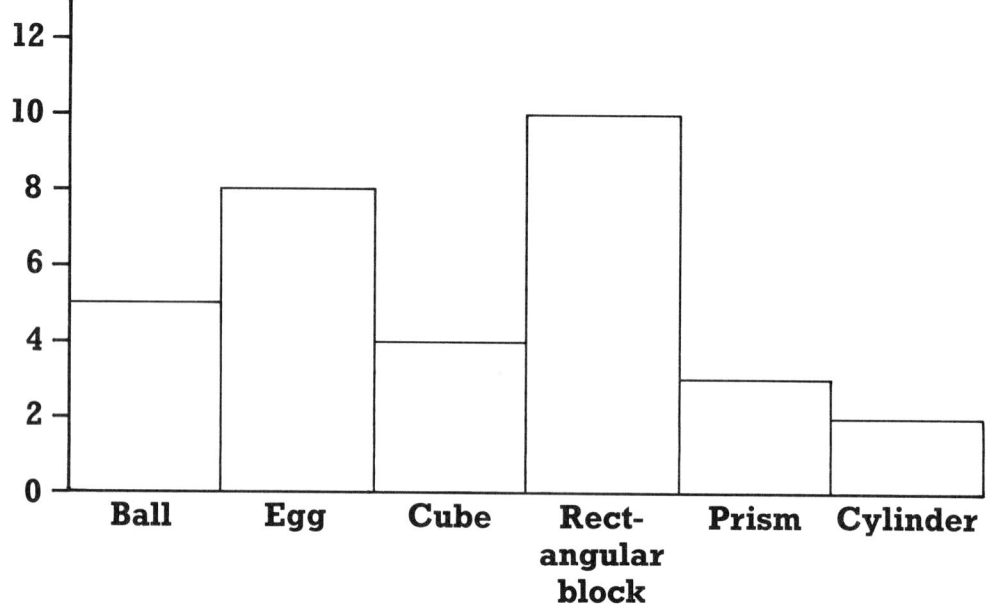

Shapes

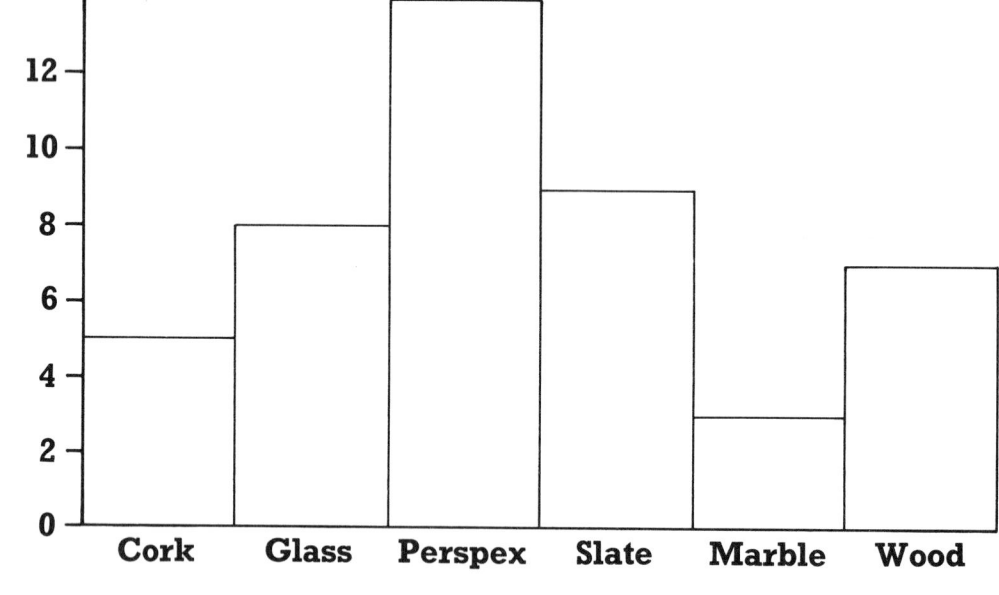

Non-metals

F S I

Making-up faces

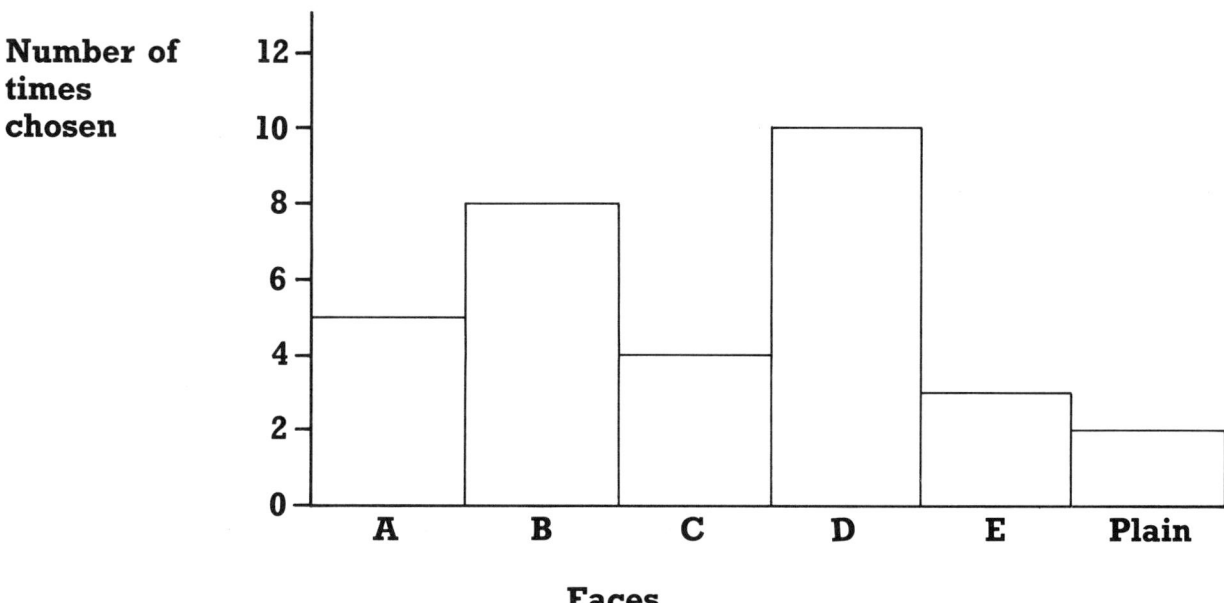

The bars can be patterned or coloured in with the same make-up colours as the faces.

Common scents

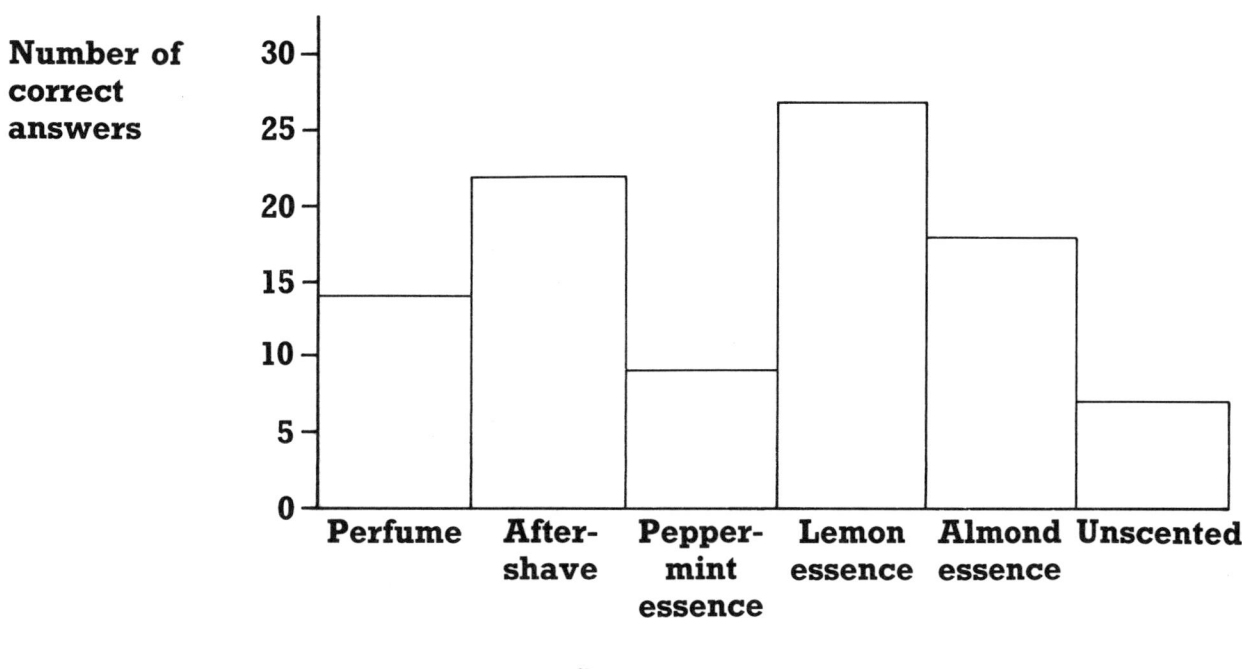

F S I

A sweet choice

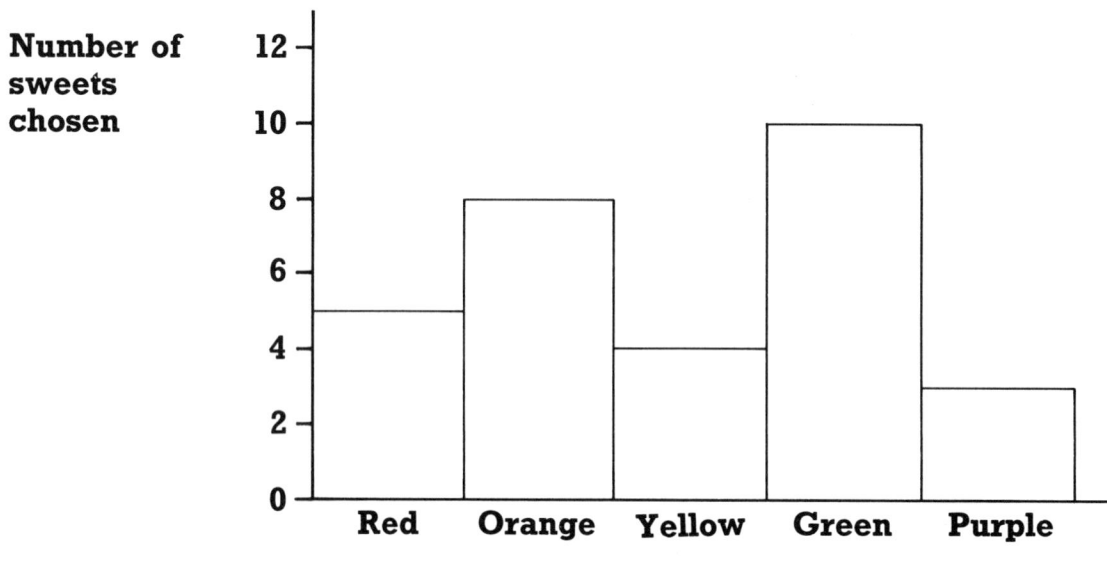

Number of sweets chosen

Colour of sweets

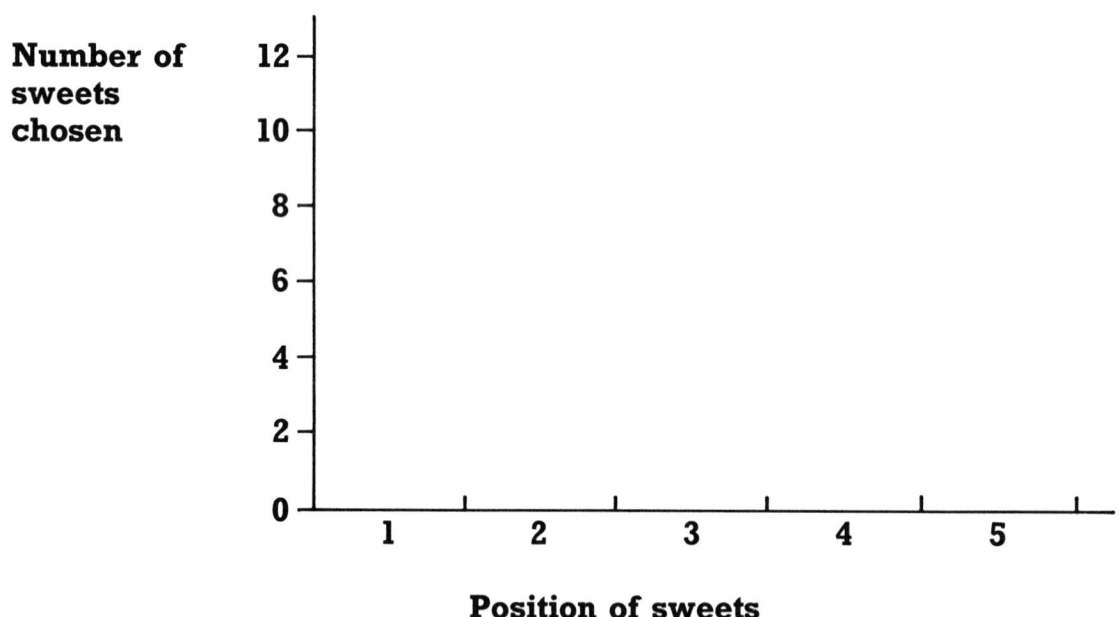

Number of sweets chosen

Position of sweets

© N.G. Furnell, 1988. Blackwell Education, Oxford

FSI

Graphs

Best bubbles

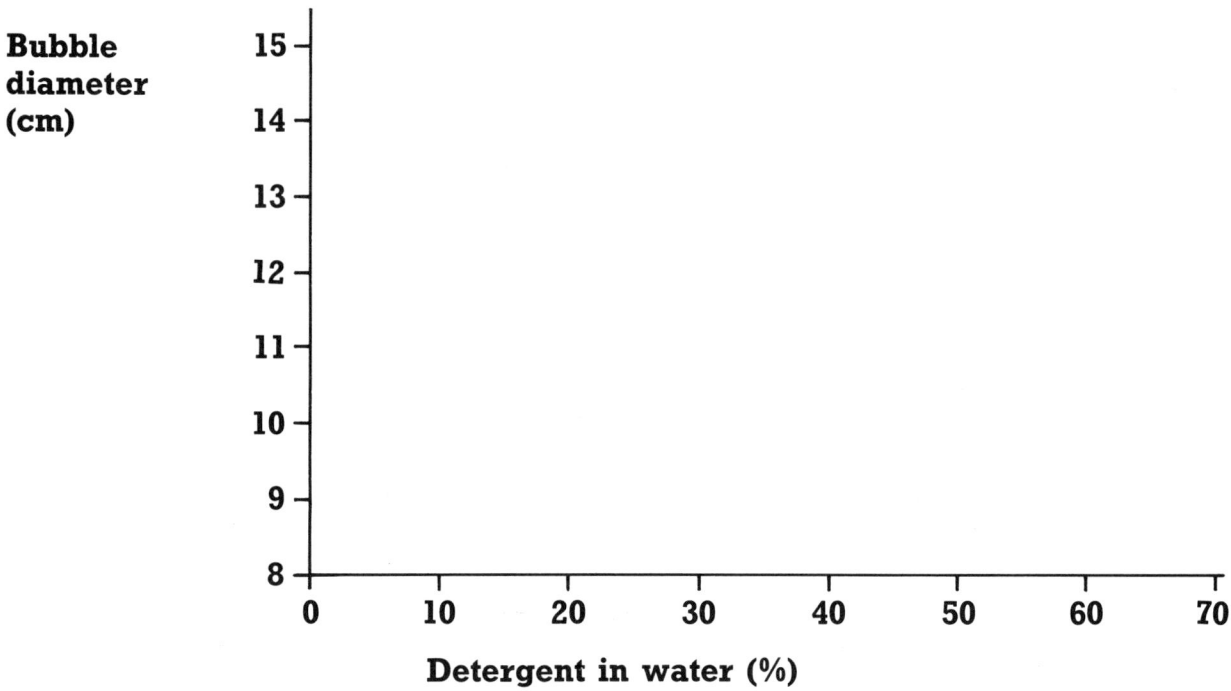

The values shown may have to be altered depending on the brand used. For example, a cheaper brand of washing-up liquid will probably need higher strength solutions than a concentrated one.

Bike stoppers!

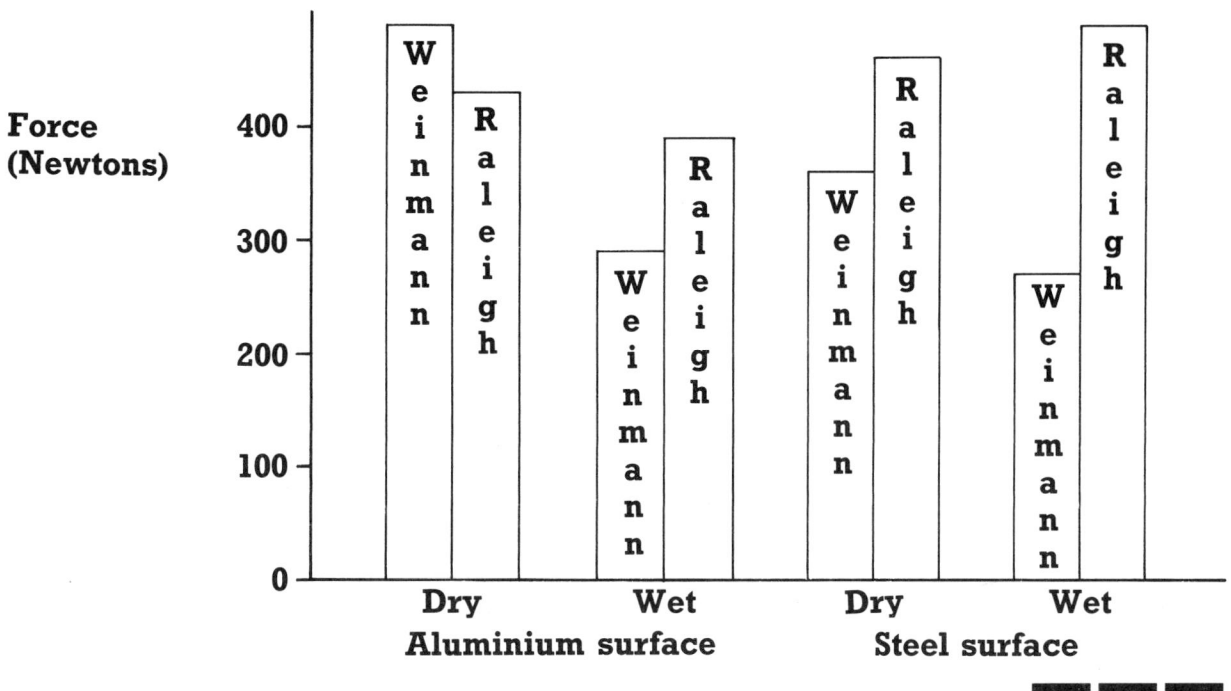

Cooling down – covering up

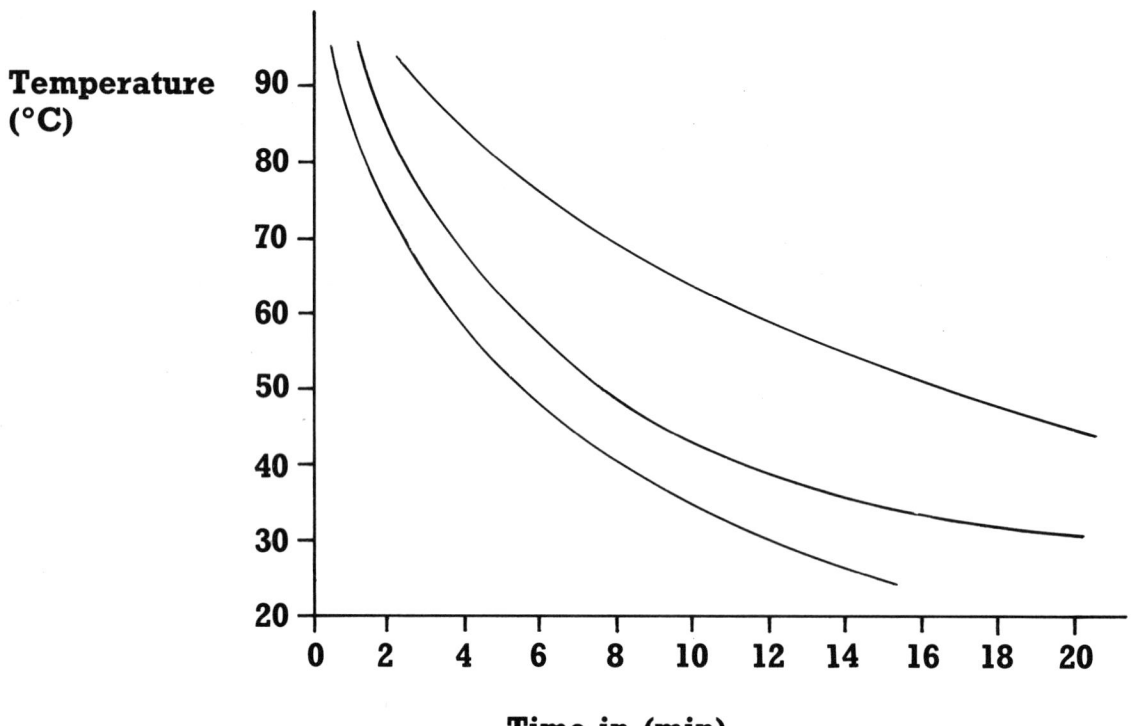

Temperature (°C) vs Time in (min)

F S I